THE MIND AND THE BRAIN

THE MIND AND THE BRAIN

A MULTI-ASPECT INTERPRETATION

by

JACK H. ORNSTEIN

MARTINUS NIJHOFF/THE HAGUE/1972

PRINTED IN BELGIUM

To all my parents

TABLE OF CONTENTS

ACKNOWLEDGMENTS

I am greatly indebted to the Canada Council for enabling me to conduct research in England during 1967-69 by means of a doctoral fellowship.

I would like to express my deepest gratitude to Avrum Stroll who first awakened my philosophical curiosity and who has provided inspiration and assistance in many of my philosophical endeavors.

My appreciation also to Stanley Malinovich and David Fate Norton who offered criticisms of an earlier draft of this book. Any errors, of course, are my sole responsibility.

I am also grateful to my typist Bonnie Wyner.

INTRODUCTION

In this book, I assess the Identity Theory of the mind and the brain. The view that the mind is the brain, or that sensations are nothing but brain processes, is an increasingly popular candidate for a solution to the mind-brain problem—the puzzle of how occurrent mental processes are related to certain cerebral processes. I concentrate here on the relationship between sensations, particularly pains, and cerebral processes and mention only briefly the subject of thoughts and other mental processes.

In an age when some researchers are trying to duplicate human thought and behavior by mechanical means, there seems to be a temptation to try to reduce human beings to the machines which they are creating. Instead of resisting this sort of nonsense in terms of antiquated talk of a human 'soul' or 'essence', I have met the Identity theorists on their own ground, namely, that of neurophysiology and scientific theory. The crux of my argument is that we can grant every empirical claim of the Identity Theory and yet deny its conclusion—that sensations are nothing but brain processes.

An historical introduction to the problem is undertaken by a discussion of Descartes' causal, interactionist view of the mind and the body. This is followed by a scrutiny of Gilbert Ryle's dispositional analysis of the mind. Ryle correctly argues that behavior can be both physical and mental but he dismisses occurrent mental states.

Using the example of being in pain as a mental state, the Identity Theory is examined first as an empirical theory and secondly as a conceptual proposal.

It is argued that the Identity Theory constitutes neither an empirical, scientific solution nor a logical solution to the mind-brain problem. The Identity Theory as an empirical hypothesis is criticized. The issue of reduction in the sciences is discussed and it is shown that Identity theorists have misconstrued the notion of scientific explanation. It is shown that no predictions or experimental findings based on the Identity Theory differ from those based on mind-brain Parallelism or Epiphenomenalism, i.e., Dualism in general. The Identity Theory, therefore, must stand or fall on its reputed conceptual advantages over Dualism.

Then the conceptual issues at stake in the mind-brain problem are discussed. The kernel of truth present in the Identity Theory is shown to be obscured by all the talk about *reducing* sensations to neural processes.

An attempt is made to characterize pain adequately as a pattern or complex of bodily processes. This view is then reconciled with the asymmetry in the way one is aware of one's own pains and the way in which others are. This asymmetry constitutes an epistemological dualism which no philosophical theory or scientific experiment could alter. The sense in which experiences are both mental and physical is thus elucidated.

A Multi-Aspect Theory of the mind is presented and defended. Five aspects of pain are discussed in some detail: experiential, neural, bodily, behavioral and verbal. Having a mind characteristically involves having all of these features except the bodily (i.e., a physical irregularity). Thus having a mind characteristically entails having experiences and a healthy, functioning brain. It also involves being able to act and speak reasonably intelligently.

From an examination of some of the different sorts of pain and the different kinds of things which can be said about them, it is concluded that no artificial language which avoids referring to psychological states would be adequate for the purposes of either everyday communication or of neurophysiological experiments.

It is concluded that the Identity Theory is of no use to science or philosophy. If followed to its logical conclusion, it ends up denying the reality of everything which is *composed* of anything, that is, it ends up denying the reality of all but the ultimate particles—unless these too are discovered to have constituents.

A Multi-Aspect Theory of mind, on the other hand, accounts for the asymmetry between self-knowledge and knowledge of others regarding experiences and it is compatible with the methods of recent neurophysiological experiments. It accounts, that is, for our ordinary notions of privacy and privileged access and for the need of a dualistic terminology. Man can be a conscious organism without being a physico-chemical mechanism.

CHAPTER ONE

DESCARTES—THE MIND AND THE BODY

Descartes is often called the father of modern philosophy. One reason for this nomenclature is that he set many of the epistemological problems which subsequent philosophers have since been trying to solve. One such problem which he fathered, or at least crystallized as no one else had done, is that of the nature of the human mind and its relation to the body.

Employing his method of doubting everything which could conceivably be false, in his search for something that was "certain and indubitable",[1] he determines that the only thing which qualifies is "the definite conclusion that this proposition: I am, I exist, is necessarily true each time that I pronounce it, or that I mentally conceive it".[2] Once assured of his own existence, which he never really doubted in any case,[3] he tries to answer the question "What am I?". He concludes that he is "not more than a thing which thinks, that is to say a mind or a soul, or an understanding, or a reason...."[4]

Descartes here equates the mind (himself) with a soul, an understanding or reason. He also at times equates it (himself) with consciousness:

But the greater part of our motions do not depend on the mind [i.e., consciousness] at all. Such are the beating of the heart, the digestion of our food, nutrition, respiration when we are asleep....[5]

Notice that these are characteristically involuntary bodily processes of which we are not normally *conscious*. They are taken care of by our

[1] *The Philosophical Works of Descartes,* ed. by E.S. Haldane and G.R.T. Ross (2 Vols.), I, 149.
[2] *Ibid.,* p. 150.
[3] *Ibid.,* p. 448.
[4] *Ibid.,* p. 152.
[5] *Ibid.,* II, 103.

autonomic nervous systems. This would seem to indicate that Descartes, at least some of the time, tends to identify the mind or mentality with consciousness. If this should be doubted, then we could say that for Descartes wherever there was a mind there had to be consciousness; there could be no unconscious mental states and no mental activities of which one were unaware.

Thus Descartes tends to identify the mind or mentality with consciousness. He also identifies himself with his mind—he is, he says, nothing but "a thing which thinks"—not with his body. But he realizes that he also is, or at least has, a *body*. And this poses a serious problem for him. According to Descartes:

the whole nature of the mind consists in thinking, while the whole nature of the body consists in being an extended thing, and...there is nothing at all common to thought and extension.[6]

Since he conceives of mind as being necessarily unextended and of body as being necessarily extended, he is forced either to deny their union or assert it and attempt to account for it. That there is such a union he is certain:

It may be concluded also that a certain body is more closely united to our mind than any other, from the fact that pain and other of our sensations occur without our foreseeing them; and that mind is conscious that these....pertain to it....only in so far as it is united to another thing, extended and mobile, which is called the human body.[7]

It is important to notice what Descartes says here about sensations, i.e., that they pertain to the mind only in so far as it is united to the body. He says that appetites, emotions and sensations "should be attributed neither to mind nor body alone, but to the close and intimate union that exists between the body and mind".[8] In other words, sensations bridge the gap between the mind and the body. When we come to the detailed discussion on pain, we shall see in what sense it is both mental and physical, i.e., in what sense it bridges the mental-physical gap.

Descartes affirms that the union of mind and body is a peculiarly intimate one:

nature also teaches me by these sensations of pain....etc., that I am not only lodged in my body as a pilot in a vessel, but that I am very closely united to it, and so

[6] *Ibid.*, p. 212.
[7] *Ibid.*, I, 255.
[8] *Ibid.*, p. 238.

to speak so intermingled with it that I seem to compose with it one whole. For if that were not the case, when my body is hurt, I, who am merely a thinking thing, should not feel pain....[9]

Descartes declares that man is "a composite entity....who consists of soul and body".[10] But how could a spiritual (unextended) substance, the soul or mind, be joined in any way to an extended substance, the body? Descartes locates the place of union in what came to be known as the pineal gland of the brain:

the part of the body in which the soul exercises its functions immediately is in nowise the heart, nor the whole of the brain, but merely the most inward of all its parts, to wit, a certain very small gland which is situated in the middle of its substance and so suspended above the duct whereby the animal spirits in its anterior cavities have communication with those in the posterior, that the slightest movements which take place in it may alter very greatly the course of these spirits; and reciprocally that the smallest changes which occur in the course of the spirits may do much to change the movements of this gland.[11]

He claims that the meeting place of mind and body, or the "seat of the soul" as he sometimes calls it, is the pineal gland. But how is it possible for two utterly dissimilar substances to interact? Would not some part of the mind have to (somehow) be *located* in the pineal gland, at least temporarily? But this, Descartes holds, is impossible, since "the mind is entirely indivisible" [12] and unextended.

The best that Descartes can do is to assert that the pineal gland acts "immediately upon the soul" [13] though the "animal spirits" seem to play a crucial role in the interaction. The mind, by some unnamed process, acts directly on the pineal gland and then the animal spirits take over. These animal spirits, which are formed from "very subtle parts of the blood", are "material bodies" whose "one peculiarity is that they are bodies of extreme minuteness and....move very quickly like the particles of the flame which issues from a torch." [14] The pineal gland, says Descartes, is "capable of being thrust to one side by the soul, and to the other by the animal spirits" [15] and they play an essential role whenever the mind acts on the body or the latter acts on the former.

---

[9] *Ibid.*, p. 192.
[10] *Ibid.*, p. 437.
[11] *Ibid.*, pp. 345-46.
[12] *Ibid.*, p. 196.
[13] *Ibid.*, p. 138.
[14] *Ibid.*, pp. 335-36.
[15] *Ibid.*, p. 353.

Many of Descartes' contemporaries were puzzled by his account of the nature of the mind and of its connection with the body. In his second letter to Princess Elizabeth, Descartes in desperation suggests that:

....by relying exclusively on the activities and concerns of ordinary life, and by abstaining from metaphysical meditation....we can learn to apprehend the union of soul and body.[16]

In other words, if one tries not to think about the puzzle, the union of mind and body will become clearer.

But critics such as Pierre Gassendi raised many serious doubts about Descartes' whole enterprise. In his objections to the sixth of Descartes' *Meditations,* Gassendi poses numerous questions about this supposed interaction between two reputedly disparate substances:

....it still remains to be explained, how that *union and apparent intermingling*....can be found in you if you are incorporeal, unextended and indivisible. For if you are not greater than a point, how can you be united with the entire body which is of such great magnitude? How, at least, can you be united with the brain, or some minute part of it, which....must yet have some magnitude or extension, however small it be? If you are wholly without parts, how can you mix or appear to mix with its minute subdivisions? For there is no mixture unless each of the things to be mixed has parts that can mix with one another. Further, if you are discrete, how could you be involved with and form one thing along with matter itself?....what must the union of the corporeal with the incorporeal be thought to be?....ought not that union to take place by means of the closest contact? But how....can that take place, apart from body? How will that which is corporeal seize upon that which is incorporeal, so to hold it conjoined with itself, or how will the incorporeal grasp the corporeal, so as reciprocally to keep it bound to itself, if in it, the incorporeal, there is nothing which it can use to grasp the other, or by which it can be grasped.[17]

Gassendi, often referring to Descartes in such terms as "my worthy Mind" or "O unextended thing", argues that it is self-contradictory to assert both that one is totally unextended and yet "diffused through the whole body".[18] Gassendi also asserts that Descartes, in denying that his body is an essential part of himself, is *ipso facto* denying that he is "a complete human being" and including merely "an inner and more hidden part...." [19] In other words, Descartes is in error the moment he identifies himself with his mind and says that his body is not an essential part of

---

[16] Cited by Norman Kemp Smith, *New Studies in the Philosophy of Descartes,* p. 258.

[17] *Descartes,* II, 201-02.

[18] *Ibid.,* p. 198.

[19] *Ibid.,* p. 138.

himself. If Descartes really were nothing but an unextended mind, then it is impossible to see how he could also be, or have, an extended body.

It is impossible to conceive how Descartes, if he were really only an unextended thinking substance, could be united, and in close causal connection, with a necessarily extended substance, the body. Yet, as we saw, Descartes insists that man is a "composite entity" consisting of both soul and body.[20]

How could this be if, as Descartes says, "not only are their natures different but even in some respects contrary to one another"?[21] He is committed to keeping mind and body apart but feels compelled by the facts of experience both to admit, and attempt to account for, their union and interaction. For example, he tries at length and in great detail in *The Passions of the Soul* to explain physiologically what he seems to rule out logically, i.e., how the mind and body can and do interact.

He never reconciles his extreme form of dualism (mind versus matter) with what he admits to be the facts of experience, namely that in almost all of our waking lives our mind and our body seem to be inextricably bound together. Indeed, as we have seen already, Descartes realizes the extent to which the mind and the *brain* are interdependent:

....although the mind of man informs the whole body, it yet has its principal seat in the brain, and it is there that it not only understands and imagines, but also perceives....the soul or mind....is intimately connected with the brain....[22]
....the soul feels those things that affect the body....only in so far as it is in the brain....For....there are many maladies which, though they affect the brain alone, yet either disorder or altogether take away from us the use of our senses....[23]

Thus Descartes realizes the importance of the brain to mentality.

His extreme dualism outlasted all his attempts to explain how two such heterogeneous substances could be united or causally interrelated. The problem still exercises some neurophysiologists, as will be seen in a later chapter.

Two of Descartes' greatest successors, Spinoza and Leibniz, accept a strict dichotomy between body and mind, but unlike Descartes, deny that mental and bodily events are causally interrelated.

In the preface to Part V of *The Ethics*, Spinoza criticizes Descartes' claims about mind-body interaction. After describing Descartes' notions

---

[20] *Ibid.*, I, 437.
[21] *Ibid.*, p. 141.
[22] *Ibid.*, p. 289.
[23] *Ibid.*, p. 293.

regarding the pineal gland as the seat of the soul and the purported role of the animal spirits in the interaction, Spinoza asks:

What does he understand....by the union of the mind and the body? What clear and distinct conception has he got of thought in most intimate union with a certain particle of extended matter? Truly I should like him to explain this union through its proximate cause. But he had so distinct a conception of mind being distinct from body, that he could not assign any particular cause of the union between the two, or of the mind itself, but was obliged to have recourse to the cause of the whole universe, that is to God. Further, I should much like to know, what degree of motion the mind can impart to this pineal gland, and with what force can it hold it suspended? [24]

These remarks echo Gassendi's doubts about the intelligibility of Descartes' conception of the mind and the body and of their relationship. Spinoza himself, though he too conceives of mind as thought and of body as extension, holds an entirely different view from Descartes regarding their connection.

He holds that there is a complete parallelism between mind (thought) and body (extension) and that there could be no causal relationship between them. How indeed, could mind and body interact if they are not different substances but one and the same thing? Spinoza holds the following view: "....substance thinking and substance extended are one and the same substance, comprehended now through one attribute, now through the other".[25]

Mind and body, according to Spinoza, are not different substances, entities or realms but are "one and the same individual conceived now under the attribute of thought, now under the attribute of extension...." [26]

In the Proof of Proposition II, Part III of *The Ethics,* Spinoza explicitly denies that there *could* be any interaction, causal or otherwise, between the mind and the body. Mind and body are "one and same thing" and it is nonsense to assert that a "physical action has its origin in the mind".[27] He adds that a "mental decision and a bodily appetite....are simultaneous, or rather are one and the same thing".[28]

We shall assess the merits of this view in the discussion of the *Dual Aspect* interpretation of the mind-body relationship.

---

[24] *The Chief Works of Benedict De Spinoza,* ed. by R.H.M. Elwes (2 Vols.), II, 246.

[25] *Ibid.,* p. 86.

[26] *Ibid.,* p. 102.

[27] *Ibid.,* pp. 131-32.

[28] *Ibid.,* p. 134.

Leibniz also denies that there is any causal interaction between the mind and the body.

The soul follows its own laws, and the body likewise follows its own laws; and they agree with each other in virtue of the pre-established harmony between all substances, since they are all representations of one and the same universe.[29]

Such a view of the relation between mind and body is today not a "live option". We regard it as more than a mere coincidence or a divinely pre-established harmony that, under normal circumstances, when I receive a serious injury I feel pain or when I try to refrain from striking someone in anger, I usually succeed. It would seem that there is an undeniable connection of *some* sort between what are called 'mental' events or processes and what are called 'physical' events or processes. This is true both at the macro-level of behavior (mind and body) and at the micro-level of neurophysiology (mind and brain).

In his recent book *The Concept of Mind,* Gilbert Ryle criticizes Descartes' conception of the mind and its relation to the body. Basing many of his arguments on ideas developed by Ludwig Wittgenstein, he argues persuasively that the mind is not a separate substance, property, entity or realm. He also challenges the notion that the mind is something wholly "inner" (spiritual) and private, arguing that the mind is not a substance *like* the body (except that it is invisible) but that the mind is in a different logical category from that of the body. This is how he characterizes the mind:

....'my mind'....signifies my ability and proneness to do certain sorts of things....[30]
....in describing the workings of a person's mind we are not describing a second set of shadowy operations. We are describing certain phases of his one career; namely we are describing the ways in which parts of his conduct are managed.[31]
To find that most people have minds (though idiots and infants in arms do not) is simply to find that they are able and prone to do certain sorts of things....[32]
....when we speak of a person's mind, we are not speaking of a second theatre of special-status incidents, but of certain ways in which some of the incidents of his one life are ordered.[33]
To talk of a person's mind....is to talk of the person's abilities, liabilities and inclinations to do and undergo certain sorts of things and of the doing and undergoing of these things in the ordinary world.[34]

[29] G. Leibniz, *The Monadology and Other Philosophical Writings,* trans. by R. Latta, p. 262.
[30] *Op. cit.,* p. 168.
[31] *Ibid.,* p. 50.
[32] *Ibid.,* p. 61.
[33] *Ibid.,* p. 167.
[34] *Ibid.,* p. 199.

According to Ryle, then, a person's mind is not a spiritual (ghostly) substance or entity inhabiting, and in some mysterious way "intermingling with", a body. A living person is not to be identified with a *mind,* as Descartes had said, but with a living, acting, thinking *body.* To say of such a being that it has a mind is to say that it can behave, is disposed to behave and does behave in certain sorts of ways, e.g., tie its own shoelaces, play chess and converse coherently with other similar beings. The concept of mind, he argues, involves both our ability and disposition to behave in certain ways and also instances of that behavior itself. ('Behavior' includes verbal behavior.) This view has come to be called a behavioral or dispositional analysis of the mind.

If Ryle has adequately characterized the mind, then the mind-body problem no longer arises. If one's mind is nothing but one's ability and disposition to behave in certain ways, and also instances of the behavior itself, it makes no sense to ask what the relationship is between one's mind and one's body. One's mind just *is* one's body, dispositionally in the sense that we are able and disposed to behave in certain ways and actually in the sense that it includes certain instances of that behavior itself. While Descartes held that the mind is spiritual, inner and private, Ryle holds that it is physical, outer and public. In the course of criticizing the Identity Theory we shall see that the truth lies somewhere between, or rather straddles, Cartesian privacy and Rylian behaviorism.

Ryle's account of the concept of mind is correct so far as it goes, but it is far from being adequate. As some of his critics have noted,[35] he is often talking more in terms of the narrower concept of intelligence than in terms of the broader one of mind or consciousness. That is, he is thinking of the mind or a mental process primarily as it is referred to in "He has a good mind" rather than as in "A wild and preposterous idea flashed through my mind" or in "The pain is excruciating". Ryle denies that sensations are mental. He claims that "we do not regard the fact that a person has a sensation as a fact about his mind"[36] but insists that "I am not....denying that there occur mental processes".[37] Here are his examples, however: "Doing long division is a mental process and so is making a joke".[38] It would seem, then, that for Ryle, sensations are not mental processes.

---

[35] Hugh R. King, "Professor Ryle and The Concept of Mind", *Journal of Philosophy,* Vol. 48 (1951), pp. 280-96.

[36] Ryle, *Mind,* p. 222.

[37] *Ibid.,* p. 22.

[38] *Ibid.*

In the most common sense of the word 'sensation' or the word 'experience', however, one could not have an *unconscious* sensation or experience. If consciousness is a criterion of mentality, which it is, then any conscious state or process must be, at least in part, a mental state or process. Pains are experiences and cannot sensibly be said to exist unnoticed; they cannot exist except as aspects of consciousness. Sensations or experiences are, at least in part, mental processes.

I am not claiming that any bodily process or entity of which we can be conscious is *ipso facto* a mental process. For example, though one may be conscious of a bruise on one's arm, it would be ludicrous to suggest that the bruise is thereby a mental process. It is otherwise, however, with sensations. One could be unaware of one's bruise but not unaware of one's sensation. Any process, bodily or otherwise, which could not in principle exist, except as an aspect of consciousness, is itself a mental process in so far as consciousness is a mental process. It would be a contradiction to say that someone is experiencing pain (i.e. feels pain) but is unconscious of the pain (i.e. does not feel pain). This, then, is one criterion of mentality for at least this class of mental processes (sensations), i.e., that they cannot exist—it makes no sense to say that they exist—unless the person having them is conscious of them. (Nobody would mind having pains he could not feel.)

It may be objected that the contradiction between 'feeling pain' and 'being unaware of (not feeling) the pain' arises because of the word 'feel' and not because of the word 'pain'. But pain is a sort of experience —a characteristically distressing one—and an experience is something which we *live through*. We could say that the contradiction arises from the word 'awareness', e.g., "He is experiencing pain [and so must be aware of it] but his attention has not been called to this fact [he is unaware of it]". Whether we call pain a feeling, a sensation or an experience—and these are not necessarily mutually exclusive—it cannot sensibly be said to exist unnoticed. Even if pain has neural correlates or is constituted by neural firings, it is still at least in part an experiential state, i.e., a mental state.[39] To say that a pain exists but is being had by nobody is to say something utterly vacuous. Nothing would be *called* 'pain' unless it were an aspect of someone's consciousness.

There are at least two aspects of mind, viz., experiential, including consciousness, and behavioral, including intelligent behavior. Ryle identifies the mind with the latter and does his best to dismiss the former.

---

[39] For further discussion on the mentality of pain, see below, pp. 142 ff.

It is my contention that there are aspects of mind other than behavior and that no theory of mind can be adequate which denies or ignores any of them.

That Ryle very often does restrict himself to intelligence, rather than deal with mentality as a whole, can be seen in the following:

Having a sensation is not an exercise of a quality of intellect or character. Hence we are not too proud to concede sensations to reptiles.[40]

True, one need not be intelligent to have sensations, but this does not mean that sensations are nothing but physical processes. In so much as an organism must be aware or conscious before it can be sensibly said to be having a sensation, we cannot deny it some degree of mentality. Though we may be reluctant to say that any being other than man has a *mind,* we know that there are varying degrees of mentality among animals. Sentience may indicate but a modicum of mentality but this does not mean that it involves nothing but *physical* processes.

J. J. C. Smart, a staunch advocate of the Identity Theory, also identifies the mind with intelligent behavior: "To say that we have a mind is to say that we behave intelligently, not that we have a soul or 'ghost'."[41] That such a view distorts our conception of the mind can be seen by reflecting upon the following locutions:

1. "Who knows what he has in mind?"
2. "Nobody could tell what he was thinking."
3. "I'd give a million dollars to know what is going on in his mind." (Or "in his head.")
4. "Do animals have minds?"

Some quotes from *The Devils* by Dostoevsky (translated by David Magarshack):

5. "Of course I didn't know what was going on inside him, I merely saw him from the outside."[42]
6. "I found him in a most remarkable state of mind: upset and greatly agitated, but at the same time decidedly triumphant."[43]
7. "A wild and preposterous idea flashed through my mind."[44]

---

[40] Ryle, *Mind,* p. 204.
[41] J.J.C. Smart, *Philosophy and Scientific Realism,* p. 106.
[42] *Op. cit.,* p. 213.
[43] *Ibid.,* p. 425.
[44] *Ibid.,* p. 428.

It is well known that people do not always disclose their thoughts to others, e.g. in statements 1, 2, 3, and 5. What is disclosed or kept concealed here is not behavior, intelligent or unintelligent, but ideas or thoughts. In regard to statement 4, we may know that some animals behave intelligently and yet be undecided as to whether we should ascribe a mind to them. A mind, therefore, does not consist of intelligent behavior alone. If we were convinced that another species of animals besides man were *conscious,* we would not hesitate to ascribe a mind to them. Descartes may have been mistaken to identify the mind with consciousness but where there could be no consciousness, we would refuse to acknowledge the presence of a mind. A robot might engage in the most sophisticated sorts of behavior and yet be mindless.

In statement 6, one might be tempted to argue that it is primarily behavior to which reference is being made. After all, explicit reference is made to the person's being upset, agitated and yet triumphant. However, Dostoevsky's express use of the locution "state of mind" is not completely gratuitous. One could conceivably produce a robot which exhibited behavior characteristic of agitation and triumph, yet if the robot were not also *feeling* agitated and triumphant, we would not call his (its) state one of *mind*. What does not have a mind could hardly be said, strictly speaking, to be in a particular state of mind.

Finally, in sentence 7, no behavior, or intelligent behavior, is in question. An idea can occur to someone utterly incapable of either expressing it or even informing others of its occurrence, e.g., someone who is paralyzed but who is still conscious. It is clear in such a case that any sort of behavior is impossible, unless one wishes to call the occurrence of an idea to someone, or of consciousness in general, a case of 'covert' behavior.

Should one insist on doing this, I suggest that our present distinction between behavior and non-behavior would merely have undergone a change in name, namely, 'overt behavior' and 'covert behavior'. In other words, the fact still remains that a person can be conscious, e.g., have ideas—undergo mental processes—and yet engage in no overt behavior at all. Mentality, therefore, cannot simply be identified with potential and actual intelligent behavior.

Though Ryle does not do complete justice to the concept of mind, he does, I think, show once and for all that the mind is not a spiritual substance or Ego inhabiting an alien body, that it is not a ghost in a machine. His dispositional analysis of the mind may not be adequate but it does account for an important aspect of mind, left out by Descartes.

As Ryle shows, the problems of other minds and solipsism (which are actually two sides of the same coin), do not arise once we see that persons are not ghostly inhabitants of their bodies but are conscious organisms. Ryle does not hold that men are automata:

Men are not machines, not even ghost-ridden machines. They are men—a tautology which is sometimes worth remembering.[45]

As we shall see, many Identity theorists do not seem to remember this. J.J.C. Smart, for example, explicitly claims that we are mechanisms.

As we said earlier, Ryle claims that the concepts of mind and matter, or mind and body, are in different logical categories. A person, according to Ryle, is not composed of both a mind *and* a body, or of mind *and* matter. Persons are material bodies which are said to have minds if they can do certain sorts of things. He sometimes says that having a mind also means that one is able and inclined to *undergo* certain sorts of things [46] but he emphasizes again and again that behavior is the key to our understanding of the concept of mind. It is clear that Ryle is mainly concerned with the mind-body problem—"I aim at nothing more than [to] rectify the logic of mental-*conduct* concepts"[47]— and not with the mind-brain problem. In other words, he is dealing with the problem of the nature of the mind and its relationship to our whole body and our behavior, not with the relationship of mental processes to the brain or brain processes. He does, however, make the following remarks about the mind-matter (and thus about the mind-brain) problem:

....the dogma of the Ghost in the Machine....maintains that....there occur physical processes *and* mental processes...but...the phrase 'there occur mental processes' does not mean the same sort of thing as 'there occur physical processes'....it makes no sense to conjoin or disjoin the two....The belief that there is a polar opposition between Mind and Matter is the belief that they are terms of the same logical type.... both Idealism and Materialism are answers to an improper question. The 'reduction' of the material world to mental states and processes, as well as the 'reduction' of mental states and processes to physical states and processes, presuppose the legitimacy of the disjunction 'Either there exist minds or there exist bodies (but not both)'. It would be like saying, 'Either she bought a left-hand glove and a right-hand glove or she bought a pair of gloves (but not both)'.[48]

There is a sense in which Ryle is right here but there is a sense in which he is, it seems to me, mistaken. He is right in saying that neither

[45] *Ibid.,* p. 81.
[46] *Ibid.,* p. 199.
[47] *Ibid.,* p. 16, emphasis mine.
[48] *Ibid.,* pp. 22-23, emphasis mine.

physical processes nor mental processes can be reduced to the other. But the reason for this is not the one which he gives, viz., that mental and physical processes are in different logical categories. Some aspects of mind may indeed be in a different logical category from that of body or matter, for example *intelligence*. To say that someone is clever is to *assess* his performances, not to describe additional, invisible ones which are observable only to himself. But there are other sorts of mental phenomena of which it is not absurd, or false, to say that they are in the same logical category as physical processes.

Let us suppose that someone is in excruciating pain. Besides the physical (including neural) events transpiring in his body, there is his feeling of pain. Without his feeling this pain, it would entirely lack sense to say of him that he is in excruciating pain. It would be absurd to say that someone is in extreme pain but is not *feeling* any pain (or 'his pain'). It seems to me that we can legitimately say, indeed must say, of a person in excruciating pain that there is occurring in him both a physical *and* a mental process. What is meant here is that the process occurring in him or the state he is in, i.e., his being in pain has both mental and physical features. There is his wound, if he has one, his brain processes and his feeling of pain. It is this last item—the sensation or feeling to which we normally refer when we say of someone that he has an excruciating pain. We give such a person a sedative to relieve him of his feeling of pain, not primarily, or even knowingly, to eliminate certain of his neural processes which are associated with, and are claimed by some to be identical with, his feeling of pain. It may be the case that eliminating the experience or feeling of pain invariably involves eliminating certain neural processes also. But whether the experiential feature of pain is eliminated by a drug, by an operation or by hypnotic suggestion, it is evident that this is the only feature with which we are ordinarily acquainted and which we wish to eliminate.

The feeling or sensation of pain is thus in the same logical category as the physiological processes in the sense that it is an occurrent state or process as opposed to being a disposition or set of dispositions. Ryle identifies mentality with dispositions to behave in certain ways and with instances of that behavior. I argue that an occurrent state or process which is not behavioral may be mental too, e.g., pains.[49]

One may want to place the feeling of pain in a different logical category from that of physical processes because of the different senses in which

---

[49] Cf. below, pp. 142 ff.

these can be 'observed'. One may wish to do so because feelings and physical processes are, on the face of it, utterly unlike. But whether we call them disparate processes or disparate aspects, we cannot plausibly deny that an occurrence with both neural and experiential features, and perhaps other features as well, is involved when someone is in pain. Whatever theory of the mind-brain relationship we adopt, we cannot sensibly deny the existence of the feeling of pain itself. If this is what some Identity theorists wish to do, then they cannot be talking about the same problem we are, viz., the status of, and relationship between, sensations of pain and certain neural processes.

The word 'pain' is the name of a feeling or sensation which we characteristically wish to avoid. There are other sorts of feelings which we would not call 'pains' but I do not think that any advantage would be gained by discussing feelings in general here. I have chosen one specific sort of sensation, feeling or experience, viz., pain, because it is the example used most frequently in the philosophical literature on the Identity Theory. So long as it is admitted that pain is a feeling or sensation—whether we characteristically seek it or avoid it—I think it is a suitable subject in a discussion of a theory which identifies sensations with brain processes.

Ryle, it seems, commits himself to the view that mental processes are really physical processes, either behavioral or physiological.[50] If someone is in agonizing pain, then on Ryle's view certain behavioral or physiological processes are occurring, but that is all. It looks as though Ryle implicitly holds some version of materialism, i.e., the view that sensations and all other supposed mental processes are really physical processes.

Ryle specifically *repudiates* materialism but, it seems to me, must be credited with it himself. Though he nowhere explicitly defends the view, I think I can make good my claim that he holds some version of it. He says:

In its familiar, unsophisticated use, 'sensation' does not stand for an ingredient in perceptions, but for a kind of perception.[51]
In our ordinary use of them, the words 'sensation' 'feel' and 'feeling' originally signify perceptions.[52]

However, he does not seem to include pains in this class of sensations

---

[50] Ryle does not specifically discuss the status of brain processes with regard to experiences.

[51] Ryle, *Mind.*, p. 201.

[52] *Ibid.*, p. 241.

for he says that we ordinarily use the noun 'sensation' or the verb 'to feel' for "tactual and kinaesthetic perceptions.... *as well as* for localisable pains and discomforts".[53]

So far, then, Ryle is not committed to the view that pains are really physical processes, potential or actual.

But in regard to pain itself, Ryle seems to hold irreconcilable opinions. He says, on the one hand, that "being in pain is a state of mind, namely one of distress" [54] and that "A pain in my knee is a sensation that I mind having; so 'unnoticed pain' is an absurd expression...." [55] On the other hand, he says things which imply that one could have a pain and yet *not* notice it.

Headaches cannot be witnessed, but they can be noticed, and while it is improper to advise a person not to peep at his tickle, it is quite proper to advise him not to pay any heed to it.[56]

The suggestion seems to be that headaches can be, but are not of necessity, noticed, and that it is proper to advise someone not to pay any heed to his tickle or pain. Ryle repeats this in a later book, *Dilemmas,* where he says:

A person whose foot is being hurt by a tight shoe....may attend to the pain.... without thinking at all about what is causing the sensation; or he may be thinking about the shoe....without paying any heed at all to the pain....he can be so absorbed in something else that for a time he totally forgets his pain....[57]

I do not see how one can totally forget the pain that one has if, as Ryle says, 'unnoticed pain' is an absurd expression. Of course, we often say to people with aches or pains "Try not to think about it" or "If you ignore your pain, it'll go away". We also deliberately distract a child's attention, after it has hurt itself, in the hope that it will forget about its pain, or not notice it so much. But in these cases we do not suppose that there will be a presently existing feeling that one has ceased to feel —we suppose that the raw feel (the feeling) diminishes or ceases as the individual attends to something else.

It may be argued that we could forget our headaches and yet not really get rid of them. Headaches may be like troubles—they may be forgotten but do not *ipso facto* diminish or disappear.

[53] *Ibid.,* p. 200, emphasis mine.
[54] *Ibid.,* p. 221.
[55] *Ibid.,* p. 203.
[56] *Ibid.,* p. 206.
[57] *Op. cit.,* p. 58.

Suppose that a child cuts itself and as it begins to cry, we immediately distract its attention and apply a bandage. If we are successful and the child experiences no pain after being distracted, would we say that he has a pain or is in pain but does not know it (is not aware of it)? What sense attaches to this notion? How could we tell whether his 'pain' is mildly distressing or excruciating? If after a while the child experiences pain in his finger, *then* he will have a pain or be in pain. We must be able to distinguish between having pains and not having pains. If pains could exist unsensed, as do many physiological processes, then it might be the case that all of us are always in constant pain but many of us fail to realize this. I submit that such a notion is vacuous. Pains are not like troubles—a forgotten headache *has* gone away. This is what we mean by a 'forgotten' headache.[58]

Ryle seems to vacillate between the view that pains are something necessarily felt or attended to, and the view that they could somehow exist and yet be ignored. Whichever view he holds, however, I think it is evident that he denies the existence of sensations which are distinguishable from physical processes or behavior.

In regard to someone in extreme pain, Ryle would say: "To call a feeling or sensation 'acute' is to say that it is difficult not to attend to it...." [59] This statement, I think, shows conclusively that Ryle is trying to analyze pains (or sensations in general) in terms of physical (behavioral or physiological) processes alone. He does seem to be committed to some version of materialism, at least as regards sensations. His "is to say" in the quotation above seems to be functioning in this statement exactly as "means nothing more than". In other words, Ryle seems to be saying that an acute sensation is just a phenomenon to which we find it difficult not to attend. It seems to me that he may be leaving out the crucial feature of, e.g., acute pains, viz., the way they feel to the individual having them. It is *because* a pain is acute, intense, severe, excruciating, etc., that we find it difficult, if not impossible, to ignore it. We do not just happen to dislike pain or happen to find it hard to ignore. Yet this might be the case if there were no feeling involved.

Whether we call pain a sensation, a feeling or an experience (or all three of these) it is self-contradictory to say that one has a pain or is

---

[58] See below, pp. 53 ff. for a discussion on the two different senses of 'sensation', i.e., the sense in which it means an experience and the sense in which it means a purely physiological process.

[59] Ryle, *Mind,* p. 88.

in pain but is not aware of the pain. Should someone say this, either we would not agree to call his state one of pain but, say, of mild discomfort, or we would say that he is in pain alright but is not the type to complain about it. We should not simply accept his statement at face value because nobody would truthfully call his state 'painful' unless he were aware of (feeling) pain.

There are places in *The Concept of Mind* where it looks as though Ryle intends to deal with the problem of the relation between sensations and physical processes. However, he says the problem does not exist because sensations are neither inside nor outside a person [60] and he restricts the discussion to "sense impressions",[61] e.g., visual and auditory sensations, and does not deal with sensations such as pain.

As to whether a sensation is inside or outside a person who is in extreme pain—though the pain may not be inside him in the sense in which his lungs or his neurons are, i.e., though it may not be observable in the same sense as they are, he himself is aware of it (feels it) as being internal or on the surface of his skin and in that crucial sense of location, pain has a location.

Smart's contention that one's experience of pain is nothing but one's neural processes, and therefore that the experience is located where one's neural processes are, is discussed in the next chapter.

To describe the mind as an immaterial, thinking substance contingently related to a body, as Descartes does, or as nothing but potential or actual behavior, as Ryle does, is to misdescribe it. Man and his experiences involve many aspects, some contingently private, some embarrassingly apparent. Persons are neither exclusively psychical, as Descartes would have it, nor primarily physical, as Ryle maintains. They are psychophysical organisms and no theory of the mind is promising which denies either one of these features.

[60] *Ibid.*, p. 208.
[61] *Ibid.*, pp. 242-43.

CHAPTER TWO

SCIENCE AND THE IDENTITY THEORY

The mind-body problem has long ceased to be regarded as that of explaining the union of, and causal relationship between, two disparate substances, the body and the mind. Few philosophers would today defend the view that the mind is a sort of ethereal substance or entity which somehow inhabits and interacts with a body. The problem is now seen as that of explaining the intimate relationship between what at least *prima facie* seem to be two entirely different kinds of processes, viz., the neural processes occurring in our central nervous system, particularly in the brain, and the mental processes, e.g., consciousness, experiences, thoughts and sensations.

The mind-brain problem can be considered entirely apart from any consideration of behavior. The problem is that of accounting for the relationship of mental processes to the neural processes with which they are somehow inextricably associated. If it is argued that the relationship between our thoughts or decisions and our gross bodily actions is the only crucial problem here, then the following illustration may convince the critic.

Nigel Calder, editor of the *New Scientist,* reports on experiments conducted with live animal brains.[1] Isolated brains of dogs and monkeys have been kept alive on the laboratory bench for as long as twenty-four hours. In fact, the brains might have lived longer but the experiments were stopped after twenty-four hours because the brains required constant attention.

These isolated brains were sustained by a constant source of oxygenated blood. Calder says that "The fact that these isolated and perfused brains continued to function was confirmed by electroencephalography

---

[1] By medical scientists in Cleveland, Ohio.

which recorded the rhythmical electric potentials accompanying activity of the cerebral nerve circuits." [2]

The following reservations expressed by Calder about these experiments on animal brains would apply with even more trenchancy to experiments on live human brains.

If these brains are truly alive they presumably retain the supreme cerebral functions of consciousness and emotion. May not the brains, deprived of all means of expression, be suffering vivid terrors and pains?
Nobody.... can say. When results of the early experiments were first announced we asked "How far along this road does consciousness and individuality persist....? Where between the present experiments and the long-range possibility of attempts to preserve an active human mind in a bottle, should we say 'Stop'?"
We now repeat these questions.[3]

It is very likely that we shall one day be able to keep an isolated human brain alive in a laboratory. And such a brain, under certain conditions, could be said to be, or at least to have, a mind. That is, if it were conscious, we would likely ascribe a mind to it.

Let us assume that brain-mapping were to be vastly improved to the point where we could infer the level of consciousness of, or even the thoughts and sensations occurring in, such a brain. (We might call this "brain-reading" or "mind-reading".)

In such a case there would clearly be no mind-*body* problem. This is because there would be no overt actions whose relationship to thoughts or decisions constitutes the traditional mind-body problem. But the mind-*brain* problem has not been eliminated. There is still the problem of how sensations, for example, are related to the brain processes which occur simultaneously with them.

Should anyone doubt that the brain alone could constitute, or sustain, a mind (in the sense of consciousness), he should reflect upon the words of the physiologist E. D. Adrian:

If the spinal cord has been severed the brain loses all nervous connexion with the lower part of the body and the patient cannot move his limbs. Yet as long as he can breathe, the mind is unchanged. The part of the body from which it can draw sensation and which it can control may have shrunk to no more than the head and the neck and a very little of the arms, but the intelligence is no less active.... The human mind is therefore anchored to the brain and not to any other part of the nervous system or the body, and if the brain survives the mind is not impaired.[4]

---

[2] "Notes and Comments: Experiments with Live Monkey Brains", *New Scientist*, p. 319.
[3] *Ibid.*
[4] *The Physical Background of Perception*, p. 6.

In regard to an imagined experiment with a human brain *in vitro,* Norman Malcolm claims that "it is a muddle to attribute thoughts, illusions or sensations to a brain".[5] It seems to me that Malcolm is clearly mistaken here. Though it may be true that we normally and rightfully attribute mental processes, for example, thoughts and sensations to *persons* as opposed to their brains or their minds, in the experiment outlined above, the brain itself could be regarded as a person. Given a certain level of consciousness, we would not hesitate to call the brain a person and, for example, to call him by the same name he had before he was shorn of the rest of his body. Particularly if we could somehow communicate with him.[6] Such expressions as "Ewart is thinking" or "He's in pain" would be perfectly in order.

The mind-brain problem, then, need not concern the will or behavior or the relationship between them. For the purposes of this discussion, we could omit any references to bodily behavior and concentrate instead on the relationship between sensations, particularly pain, and brain processes.

Ryle and other behaviorists have erred in characterizing the mind as exclusively behavior, potential and actual. There is another and crucial aspect of mind which they leave out entirely, viz., such phenomena as sensations and consciousness. Ewart, the living brain, may not be able to behave at all but he could be conscious and undergo experiences, for example, sensations. If one's having a mind consisted exclusively in potential and actual behavior, there would be no mind-brain problem. But 'intelligence' and other mental ascriptions referring largely, if not exclusively, to behavioral dispositions are only one aspect, albeit an important one, of the mind. Consciousness and sensations are another crucial feature of mentality and it is the latter, particularly pains and their connection, or alleged identity, with brain processes that is under scrutiny here.

The Identity Theory is the view that consciousness or sensations are really processes occurring in the brain. They are not composed of "mind-stuff" or of mental or thinking substance as Descartes had maintained but of electro-chemical processes in the brain. Sensations, on this view, are strictly *identical* with brain processes.

---

[5] "Scientific Materialism and the Identity Theory", Abstract, *The Journal of Philosophy,* p. 663.

[6] This would be possible, given that we had perfected a detailed brain-mapping and could feed information into, and read information coming from, the brain.

The Cartesians maintained that sensations and other mental processes are utterly dissimilar to physical processes. On this view, it proved impossible to make sense of the notion that these two different types of processes could be related at all, for example, causally, especially as the Cartesian view of causality ruled out any causal relationship between utterly dissimilar processes. This view also led to unpalatable problems regarding personal identity and knowledge of other minds.

Another possible view is that sensations and brain processes just *happen* to be correlated. One version of this position is the view that mental processes and physical processes are involved in a cosmic harmony pre-established by God. Such a view is not a surprising result of Cartesianism which held both that mental and physical processes have nothing in common and that cause and effect *must* have something in common.

None of these mind-body or mind-brain theories has proved anything like adequate. All the evidence indicates that there is an intimate relationship between brain processes and sensations. Neurophysiologists and physiological psychologists have only begun to conduct detailed investigations into the workings of the brain and into mental process-brain process correlations. It seems inevitable, however, that invariable correlations will be discovered between, for example, all sensations and certain brain processes. It is necessary, therefore, to examine the relationship between, for example, the sensation of pain and the brain process invariably associated with it—to see if this is mere coincidence, a cause and an effect, an identity, or perhaps some other relationship altogether.

It has long been known that the *sine qua non* of consciousness and mentality in general is a reasonably healthy and intact brain. Descartes himself realized that the brain, or some part of it, was somehow immediately involved with mental processes, though he thought that the mind or person, being a separate substance from the body, could and did survive the death of the body. For Descartes, the criterion of personal identity was not one's body but one's mind—a person is "a thing which thinks"—and when a person's body dies, the person himself (the mind or soul) continues to live.

However, the identification of a person with his mind is extremely problematic. In his book *Individuals,* P. F. Strawson argues, I think successfully, that the "concept of person is logically prior to that of an individual consciousness" and that a person is neither an "animated body" nor an "embodied anima". "A person", he says, "is not an embodied ego, but an ego might be a disembodied person, retaining the logical benefit of individuality from *having been* a person." [7] Thus our concept of a person as a mind or consciousness which could conceivably exist apart from the brain and the rest of the body, is logically grounded in our concept of a person as a whole. In other words, we would not have been able to conceive of persons as psychical beings unless we had already conceived of them as psychophysical beings.

While conceptual analysis has shown that the concept of mind or a mental process is logically dependent upon that of a person and his body, scientific investigation has shown that the mind or a mental process is physically dependent on the brain. As early as 1890, William James could write the following in the first chapter of his book *The Principles of Psychology:*

> If the nervous communication be cut off between the brain and other parts, the experiences of those other parts are non-existent for the mind. The eye is blind, the ear deaf, the hand insensible and motionless....if the brain be injured, consciousness is abolished or altered....A blow on the head, a sudden subtraction of blood, the pressure of an apoplectic hemorrhage, may have the first effect; whilst a very few ounces of alcohol or grains of opium or hasheesh, or a whiff of chloroform or nitrous oxide gas, are sure to have the second. The delirium of fever, the altered self of insanity, all are due to foreign matters circulating through the brain, or to pathological changes in that organ's substance. The fact that the brain is the immediate bodily condition of the mental operations is indeed....universally admitted nowadays....[8]

Neurophysiologists have conducted several experiments on the brain and have succeeded in "mapping" many of its functions. That is to say, empirical research has succeeded in establishing correlations of many parts and processes of the brain with different mental processes. For example, proper functioning of the reticular formation in the brain stem is necessary if a person is to be kept in a conscious state and the cerebral cortex must be intact and functioning normally for an individual to be able to respond to a stimulus with anything more than a mere reflex reaction. We now know which parts of the brain correspond to the different senses, for example olfactory, visual and auditory, and neurophysiology seems to be rapidly approaching the time when the mechanisms of memory will be found. However, though the brain and its functions are gradually becoming less of a mystery, the mind-brain problem still troubles many neurophysiologists. They know *that* neural processes are a necessary, and perhaps sufficient condition of mental processes but

---

[7] *Loc. cit.,* p. 103, emphasis mine.
[8] *Loc. cit.,* p. 4.

they confess their ignorance as to exactly *how* these different sorts of processes are related. They do not know, and cannot even conceive, how physical processes could result in mental processes. Wittgenstein, in *Philosophical Investigations,* voices this incomprehension:

The feeling of an unbridgeable gulf between consciousness and brain-process.... This idea of a difference in kind is accompanied by slight giddiness....When does this feeling occur in the present case? It is when I, for example, turn my attention in a particular way on to my own consciousness, and, astonished, say to myself: THIS is supposed to be produced by a process in the brain!....[9]

He then adds that there is nothing paradoxical in the statement "THIS is produced by a brain-process!" since it could be sensibly uttered "in the course of an experiment whose purpose was to shew that an effect of light which I see is produced by stimulation of a particular part of the brain".[10]

True, the statement that a mental process is produced by a brain process may not in itself be a paradox. Indeed, it is an attested fact that stimulation of certain areas of the brain produces a variety of mental processes. Nevertheless, the "difference in kind", which Wittgenstein mentions, between neural processes and experiences which they produce bewilders many neurophysiologists to this day. Some neurophysiologists still talk in terms of an "unbridgeable gulf" between neural and mental processes. In criticizing the theory which simply identifies these processes, it will be argued that the apparent gulf can only be bridged by analysis and not merely by empirical conjectures and testing. Once empirical research has shown conclusively that there is an invariable correlation between, for example, certain pains and certain brain processes, the precise relationship between these phenomena can be elucidated only by analysis and not by further empirical testing.

Leibniz gives a striking example of an imaginary experiment which illustrates the incongruency felt to exist between physical and mental processes:

....it must be confessed that *perception* and that which depends upon it are *inexplicable on mechanical grounds,* that is to say, by means of figures and motions. And supposing there were a machine, so constructed as to think, feel, and have perception, it might be conceived as increased in size, while keeping the same proportions, so that one might go into it as into a mill. That being so, we should, on examining its interior, find only parts which work one upon another, and never

---

[9] *Loc. cit.,* p. 124.
[10] *Ibid.*

anything by which to explain a perception. Thus it is in a simple substance, and not in a compound or a machine, that perception must be sought for.[11]

When a neurophysiologist conducts an operation on the brain, when he stimulates parts of the brain electrically or chemically and records the pattern of the brain's electrical activity, he sees or records "figures and motions" as Leibniz puts it. He is, on the face of it, having to do with physical processes, not with mental processes.[12] The latter are reported by the subject as they occur, if he is conscious, or they may be imparted later if he is sleeping during the experiment.

The relationship between neural and mental processes has often been regarded as inconceivable rather than merely unobservable. William James, for example, expresses it this way:

Mental and physical events are, on all hands, admitted to present the strongest contrast in the entire field of being. The chasm which yawns between them is less easily bridged over by the mind than any interval we know.[13]

And this belief that we are facing an unbridgeable gap has been stated more recently by a leading neurophysiologist, Wilder Penfield:

....for the neurophysiologist there is a working boundary [between the brain and the mind] that does exist. Physiological methods bring him nearer and nearer to it. But he comes to an impasse, and beyond that impasse no present-day method can take him.[14]

We are concerned with the most recent attempt to bridge the gap between neural and mental processes—the Identity Theory. In the course of criticizing this theory, I shall elaborate and defend an alternative view of the mind-brain relationship. I shall argue for the view that sensations are identical with neural processes only in the sense in which words can be said to be identical with the letters out of which they are composed or statements can be said to be identical with the words out of which they are constituted.[15] Pains involve not simply neural processes and a feeling

---

[11] G. Leibniz, *The Monadology and Other Philosophical Writings*, pp. 227-28.

[12] When the concept of pain is discussed, it will be made clear how very unlike are its experiential and its observable aspects.

[13] *Op. cit.*, p. 134.

[14] Wilder Penfield and Lamar Roberts, *Speech and Brain-Mechanisms*, p. 8.

[15] Avrum Stroll's analysis of the relationship between statements and sentences has at least this much in common with my analysis of sensations and brain processes —we both argue that S may be *identical* with B and yet *different* from B; see his "Statements", especially pp. 198 ff.

but both of these, and more besides. Ordinary cases of pain involve a whole pattern of events and only in cases such as an isolated, living brain could it plausibly be argued that pains involve only neural processes and feelings.

Neither monistic materialism, for example the Identity Theory, nor Cartesian dualism adequately characterizes sensations or other mental processes. It seems that commitment to a metaphysical theory, whether monistic or dualistic, has more often been an obstacle than an aid to our understanding of sensations and their relationship to neural processes.

The Identity Theory postulates an identity between neural and mental processes. This identity is formulated as "mental processes are neural processes" and not as "neural processes are mental processes". This is because Identity theorists are claiming that mental processes are *really,* or are *nothing but,* neural processes. It is, therefore, a materialistic, reductionist theory. Among the philosophers who have recently espoused some version of the Identity Theory are: Herbert Feigl, Richard Rorty, Paul Feyerabend, R. J. Hirst, U. T. Place, J. J. C. Smart and D. M. Armstrong.

The Identity theorist identifies mental processes with neural ones. The theory is sometimes called "Central State Materialism". D.M. Armstrong, for example, espouses the following view in *A Materialist Theory of the Mind:*

(Central State) Materialism....identifies mental states with purely physical states of the central nervous system. If the mind is thought of as 'that which has mental states', then we can say that, on this theory, the mind is simply the central nervous system, or, less accurately but more epigrammatically, the mind is simply the brain.[16]

I shall refer to any theory of the mind-brain relationship which holds that mental processes such as experiences, consciousness or sensations are nothing but (or are 'really') neural processes as an Identity Theory. Most Identity theorists stress that their thesis is conceptual rather than empirical. They argue that their thesis is more plausible than alternative mind-brain theories because it reduces the number of entities, or the types of processes, which are postulated by the other theories. The argument that the Identity Theory should be accepted because it represents a conceptual advance over other theories will be discussed in the following chapters. In this one, the claim that the Identity Theory is at least partially scientific or empirical will be assessed.

[16] *Op. cit.,* p. 73.

What, then, is the status of the Identity Theory? U. T. Place, in a seminal paper on the Identity Theory, describes it like this:

....the statement "Consciousness is a process in the brain", although not necessarily true, is not necessarily false. "Consciousness is a process in the brain", on my view is neither self-contradictory nor self-evident; it is a reasonable scientific hypothesis, in the way that the statement "Lightning is a motion of electric charges" is a reasonable scientific hypothesis.[17]

J.J.C. Smart, one of the most vociferous Identity theorists, says the following:

....it may be the true nature of our inner experiences, as revealed by science, to be brain processes, just as to be a motion of electric charges is the true nature of lightning, what lightning really is.[18]

The Identity Theory is the view, then, that some mental processes, for example, consciousness, experiences or sensations, are to be identified with certain physical processes, for example, brain processes. Nowhere do we find Identity theorists denying that people are conscious or that they feel pains. The claim is, rather, that these mental processes are identical with or are nothing over and above brain processes, that they are nothing but, or are really, brain processes. What people all along, or at least since Descartes, thought were mental processes occurring in some ghostly or gaseous medium called "the mind" are, according to Place and Smart, really nothing but physical processes occurring in the brain.

The Identity Theory purports to be a solution to the mind-brain problem. According to Identity theorists, for example Place, most mental processes can be analyzed dispositionally, i.e., behavioristically. Those that have so far proved recalcitrant to such an analysis, for example, consciousness, experiences or sensations, are now to be analyzed solely in terms of brain processes. Thus there are no mental processes over and above behavioral and neural processes, i.e., the mind-brain problem has finally been solved. The mind, on this view, is no longer something over and above, or besides, the body—it *is* the body, or bodily processes—either potentially or actually. If this theory is accepted, it means that any sort of mind-body or mind-brain dualism must be rejected. All mental processes, on the Identity Theory, are really only physical processes. A materialistic monism is the philosophical view of Identity theorists.

---

[17] "Is Consciousness a Brain Process?", in *The Philosophy of Mind*, ed. by V.C. Chappell, p. 102.
[18] *Philosophy and Scientific Realism*, p. 93.

Though it will, hopefully, make for greater clarity to discuss the empirical and conceptual issues regarding the Identity Theory separately, it must not be thought that all its exponents consistently defend either one of these positions alone. For example, U.T. Place holds that there are conceptual or logical issues as well as empirical ones. J.J.C. Smart vacillates between the claim that the Identity Theory represents a great empirical discovery, and the claim that it has no empirical consequences different from Epiphenomenalism, i.e., that it is preferable on logical grounds alone.

Does the Identity Theory, then, qualify as an empirical, scientific theory? Before answering this question, we must see what it is which makes for a genuinely scientific theory or hypothesis.

For a statement to have empirical content, it must be testable by observation or experiment. This means that there must be some way of telling whether the state of affairs depicted by the statement either obtains or does not obtain. If there is no way of telling, by observation or experiment, whether the statement is true or false; if, that is, the statement is compatible with any conceivable state of affairs, then it is utterly devoid of empirical content. It cannot tell us anything, it cannot inform us of anything, unless its truth (or falsity) would make some detectable difference in the world, i.e., unless there were a detectable difference between its being true or false.

A statement which risks nothing, can tell us nothing. For example, the statement "There is a desk in the next room" can be tested by a simple procedure, namely, by going into the next room and looking. But such statements as "God loves us all" are characteristically held to be untestable or unfalsifiable. Any conceivable state of affairs is deemed by some to be compatible with such statements—that is why they are uninformative.

The more a statement risks, the higher is its empirical content. If one said "There is a large, green, metal desk in the next room", more risk is being taken than by the bare statement "There is a desk in the next room."

These remarks apply not only to statements but to hypotheses or theories. The mark of a scientific theory is that it is testable by observation or experiment. Those theories which risk the most, are potentially most informative, i.e., have potentially the most empirical content.

The truth, or falsity, of a genuinely empirical theory must make a detectable difference in the world. If we could not, even in principle, make any observation or devise any experiment whose outcome would tell for or against the truth of a given statement, then that statement lacks empirical content. Such a statement could convey no information about the world because any conceivable state of affairs is compatible with it. In other words, it *makes no difference* whether such a statement is true or false. Who could ever tell, i.e., *whether* such a statement were true or false? In brief, the more a statement risks or forbids, the more empirical content it has. The more it allows, i.e., the less it prohibits, the less empirical content it has. A tautology risks nothing—it has no empirical content.[19]

Has the Identity Theory empirical content? Is it testable? R.J. Hirst, in *The Problems of Perception*, specifies what would falsify the Identity Theory. He says that "If it could be proved that some mental activity occurred without any corresponding brain activity, that would be fatal to the theory as I conceive it." [20]

Undoubtedly, the theory that sensations, or other mental processes, are identical with neural processes would be falsified if sensations occurred with no corresponding brain processes occurring. The problem is, however, that such an occurrence would falsify all those theories which postulate an invariable *correlation* between sensations and brain processes. Thus the same occurrence which would falsify the Identity Theory would also falsify psychophysical dualism.

This fact leads us to suspect that the Identity Theory may have no empirical consequences different from psychophysical dualism. The question is this: Is there any way of empirically (observationally or experimentally) distinguishing between a Correlation Hypothesis and the Identity Theory? If the same state of affairs, viz., the occurrence of a sensation without any corresponding brain process, would falsify both the Identity Theory and the Correlation Theory, then it is worth investigating whether a different state of affairs would *verify* both theories.

Suppose that in all cases where sensations occur, a corresponding brain process occurs. An Identity theorist would appeal to this fact as confirming his theory. However, a psychophysical dualist could, with as much justification, appeal to this fact as confirming his dualistic theory. It appears, therefore, that there is no empirical state of affairs which would force us to pronounce either of these theories to be true, or to "hold", though we can conceive of one which would enable us to reject both of them.

[19] Cf. Karl R. Popper, *The Logic of Scientific Discovery,* especially section 6.
[20] *Op. cit.,* p. 199.

It might be argued that we can conceive of an experiment which would establish the Identity Theory and refute psychophysical dualism. Let us consider Leibniz's example cited on page 25. Suppose that we constructed a sentient machine large enough for us to enter. If the machine honestly reported a sensation and we could find nothing but neural processes, would this not confirm the Identity Theory and refute dualism? We could just as well have used the example of observers being somehow miniaturized so that they could enter the body of a living person and see his neural processes. Supposing the machine or person to undergo pain, would the Identity Theory be established if we observed nothing but neural processes occurring? I think not. This case does not differ in principle from others examined later in this book where people could observe all the neural processes occurring inside someone experiencing pain.[21]

We should not *expect* to find (see) a sensation of pain in addition to observing the neural processes. I shall be arguing throughout this discussion that experiences, including sensations, are not the sort of thing that can be observed, either by a single observer or intersubjectively.[22]

A sensation is an integrated complex of neural processes as they are occurring in a more or less conscious being. This sensation as lived through by the subject could no more be seen than could an idea or concept which the subject has. When he has an idea or a pain, we might be able to see the neural factors involved but never the idea or experience itself. But this is not because the experience would be *hidden* from us. It would be no more hidden from our sight than from the subject's sight. Phenomena such as ideas or concepts and experiences can be seen by no one, but they are no more "hidden" than is a vacuum. (Nor if we were placed in an engine which was operating, should we expect to *see* its power or energy.)

We might, then, be so placed that we could see all the neural components of pain and still not see the pain. This is because pain is a feeling or an experience. If we were somehow embedded in the brain of someone who was feeling elated, would we see his elation when we see his neural processes? If, as is the case, we would not, must we conclude that his elation is nothing but those processes which we *can* see? To theorize that nothing exists or is real but what is in principle inter-

---

[21] Cf. below, pp. 51 ff.
[22] Cf. below., p. 54.

subjectively observable and that all things are reducible to their components is to banish from existence (or reality) not only experiences and concepts, vacuums and power, but ultimately theories too.[23] There would be nothing left of them but marks on paper, i.e., their 'ultimate constituents'.

If the claim were made that a sensation had occurred without any corresponding brain activity, the Identity theorist or psychophysical dualist could always maintain that we had simply failed to detect the corresponding brain activity. If we conducted repeated experiments and found that a certain individual, X, experienced sensations with no corresponding brain activity, the critic could say that X is a liar, or is deceived, or does not know the meaning of the word 'pain'.

Indeed the critic may be right in one or the other of these claims. However, if he forbids anything to disconfirm his theory, he renders his theory unfalsifiable.

We conclude that even if the Identity Theory can be said to have empirical content since it is falsifiable, it is utterly useless as a tool for research unless it is empirically distinguishable from dualistic mind-brain theories. The problem is to conceive of some observation or devise some experiment whose outcome would enable us to reject all other hypotheses but that of Identity.

Could the Identity Theory be empirically confirmed? Suppose that neurophysiology has advanced to such a stage that it was established beyond a reasonable doubt that no sensation ever occurred without its corresponding brain processes occurring. We shall assume that there is a well established concomitant variation between every type of sensation and certain types of neural activity.

It would not be amiss at this point to consider the means which would have to be employed in order to establish such a mental process - brain process correlation. Individuals would have to undergo and report various sensations as the researchers confirmed that this or that sort of brain process were occurring. *Prima facie*, then, a correlation is being established, not an identity. The researchers, after all, are correlating brain processes with sensations.

The question which the Identity Theory poses is whether these sensations are really brain processes or are something other than, or more than, these physical processes. But since an invariable correlation could never be established between any given entity or process and *itself*, there

<hr />

[23] Cf. below, pp. 84 ff.

must be two different types of processes or two different aspects of a single process being correlated.

It would seem that the empirical findings, i.e., the invariable correlation established between certain mental and brain processes, are as compatible with at least some versions of mind-brain dualism, e.g., Parallelism, Interactionism and Epiphenomenalism, as they are with the Identity Theory. As in the case of falsification, there seems to be no empirically detectable difference between what 'confirms' the Identity Theory and what 'confirms' certain dualistic theories.

A distinguishing feature of a scientific thesis is that it can be tested independently of any other claim, i.e., it can actually be *tested*. As Ernest Nagel says, "a function of verification is to supply satisfactory evidence for *eliminating* some or all of the hypotheses we are considering".[24]

No Identity theorist has ever outlined, or even suggested, observations or experiments whose outcome would eliminate all other prevalent theories but his. Unless such an observation or experiment is proposed, it will not be possible sufficiently to differentiate the Identity Theory from other theories so that it can be independently tested. At the present time, the Identity Theory is not testable, in the sense that counts, and is therefore of no use in furthering scientific research.

Even if the Identity Theory did hold, unless there were some way in which we could *tell* that it did, it would be of no use to us at all. Unless it is somehow rendered independently testable, we cannot use the Identity Theory to predict any phenomena which we cannot now predict on the basis of a correlation hypothesis.

We shall now examine in some detail some formulations of the Identity Theory in an attempt to distinguish between their empirical and conceptual features. Would acceptance of the Identity Theory result in the discovery of new facts or would it rather provide us with a more economical and plausible description, or explanation of facts already known?

The first exposition of the Identity Theory under discussion appeared in 1956 in U.T. Place's "Is Consciousness a Brain Process?". Since that time, the theory has gained several adherents.

Place argues that the statement "Consciousness is a process in the brain" is a "reasonable scientific hypothesis in the way that the statement 'Lightning is a motion of electric charges' is a reasonable scientific

---

[24] Ernest Nagel and Morris Cohen, *An Introduction to Logic and Scientific Method*, p. 210.

hypothesis".[25] He does not wish to argue "that when we describe our dreams, fantasies and sensations we are talking about processes in our brains" and admits that statements about consciousness are not statements about brain processes.[26] He is not making the claim that both the connotations and denotations of words referring to sensations and to brain processes are synonymous or identical. This would be to claim a necessary identity between sensations and brain processes, as is the case in such analytic statements as "A square is an equilateral rectangle". This statement is true by definition whereas the assertion that sensations are brain processes, he says, is contingent and is amenable to verification by observation.

The proposed identity, then, concerns the denotation of sensation-words and brain process-words. Their denotation or referent, he says, is one and the same, i.e., physical processes occurring in the brain. Just as lightning is nothing more than, or is 'really' a motion of electric charges, so sensations are nothing more than, or 'really', brain processes. Just as lightning is both explained in terms of, and is constituted by nothing but, the motion of electric charges, so sensations are both explained in terms of, and are constituted by nothing but, brain processes. And, just as the procedures required to determine the occurrence of lightning are radically different from those required to detect the presence of a motion of electric charges, so the operations required to verify statements about sensations and statements about brain processes are radically different.[27]

Closer inspection of one's sensation, Place admits, will never reveal the neural processes themselves, but neither will a closer look reveal a motion of electric charges when we are observing lightning. The one in pain observes his sensation in a different way from that in which the neurophysiologist does, Place claims, but the sensation is in reality those same neural processes which the latter is observing.

If his account is correct, he says, the following is the case:

....in order to establish the identity of consciousness [or sensations] and certain processes in the brain, it would be necessary to show that introspective observations reported by the subject can be accounted for in terms of processes which are known to have occurred in his brain.[28]

[25] Place, "Consciousness", p. 102.
[26] Ibid.
[27] Ibid., pp. 105-106.
[28] Ibid., p. 106.

If sensations, e.g., pains can be accounted for solely in terms of neuro-physiology, then, he says, the identity of them with neural processes will have been established.

But the evidence shows that the experience of pain, for example, cannot be explained in terms of physical processes alone. Ronald Melzack, in "The Perception of Pain", cites several experiments which indicate the following:

The psychological evidence strongly supports the view of pain as a perceptual experience whose quality and intensity is influenced by the unique past history of the individual, by the meaning he gives to the pain-producing situation and by his "state of mind" at the moment. We believe that all these factors play a role in determining the actual pattern of nerve impulses ascending to the brain and travelling within the brain itself. In this way pain becomes a function of the whole individual, including his present thoughts and fears as well as his hopes for the future.[29]

Many experiments were conducted on both animals and human beings. In one experiment, about "35 per cent of the patients report marked relief from pain after being given a placebo".[30] In another, the mere appearance of the word 'pain' in a set of instructions caused subjects to "report as painful a level of electric shock they did not regard as painful when the word was absent from the instructions".[31]

The scientific evidence indicates, then, that an explanation of pain strictly in terms of physical and neural processes is bound to be inadequate. However, the Identity theorist can claim that this evidence in no way vitiates his theory. He can assume, as Melzack himself does, that:

....psychological processes such as memories of previous experience, thoughts, emotions and the focusing of attention are *in some way* functions of the higher areas of the brain—that they *represent* the *actual activities* of nerve impulses... these higher brain functions are able to modify the patterns of nerve impulses produced by an injury. Remarkable evidence for such complex neural interplay has recently been observed in physiological laboratories.[32]

Melzack assumes here that the mental processes which seem to influence the quality of our experience of pain are themselves really only the effects of the actual agents of change, neural processes. Thus the whole chain of causation from beginning to end is constituted only by neural processes. Melzack may be unable to state in what way the mental pro-

---

[29] In *Psychobiology,* ed. by *Scientific American,* p. 307.
[30] *Ibid.,* p. 301.
[31] *Ibid.*
[32] *Ibid.,* pp. 301-02, emphasis mine.

cesses are the effects of neural processes but, as we have seen, he is not the only one who confesses ignorance here.

The Identity theorist will say that what Melzack calls a "perceptual experience", i.e., pain, is itself really a neural process. In fact, there are actually no psychological processes occurring over and above the neural processes, not even as effects, or functions, of these neural processes. He will say also that there is no question here of correlating two different types of process, neural and psychological, and then trying to account for the latter in terms of the former. An Identity Theory must not be confused with Dualism. The memories, the anticipation and the sensation of pain itself are, according to the Identity Theory, all neural (or behavioral) processes.

Place would have to respond in the second way. We see now how radical the Identity Theory is. It is not to be confused with Epiphenomenalism, which acknowledges the existence of mental processes (over and above neural ones) though it holds them to be utterly inefficacious. A consistent Identity Theory is thoroughly reductionist and materialistic —it categorically denies that there are mental processes which are different from neural or behavioral processes.

In his paper, "Is Consciousness a Brain Process?", Place suggests that "Consciousness is a brain process" is a contingent statement which has to be "verified by observation".[33]

In a second paper, "Materialism As A Scientific Hypothesis", Place recognizes that it is among the "philosophical issues that the real battle will be fought". But he still maintains that "materialism can and should be treated as a straightforward scientific hypothesis".[34] Place inspired much of the recent discussion of the Identity Theory—J.J.C. Smart acknowledges that he takes his departure from Place's first paper on this subject—and in some ways he has a better grasp of the problems involved than do later adherents to the Identity Theory.

Place says that there are two aspects to the mind-brain problem, a logical one and an empirical one. He contends that "There are certain logical conditions which must be satisfied to enable us to say that a process or event observed in one way [e.g. one's experience or feeling of pain] is the *same* process or event as that observed in (or inferred from) another set of observations made under quite different conditions [e.g.,

---

[33] *Op cit.*, p. 103.
[34] In the *Philosophical Review*, p. 104.

one's neural processes]."[35] The two criteria which he suggests are :

1) "That the process or event observed in or inferred from the second set of observations should provide us with an explanation....of the very fact that such observations are made" [36] and;

2) "That the two sets of observations must refer to the same point in space and time, allowing for such things as....differences in the precision with which location is specified in the two sets of observations".[37]

Place argues that once these two criteria or, perhaps, some other criteria have been agreed upon, then we are left "with a purely empirical issue, namely, whether there is in fact a physiological process, be it in the brain, the heart....or the big toe, which satisfies the logical criteria required to establish its identity with the sensation process".[38]

He goes on to argue that we have already made enough progress in physiological research to determine 1) that it is in the brain where we will find the appropriate physical processes with which to explain (identify) consciousness, and 2) that certain parts of the brain have already been eliminated as possible loci of consciousness. Thus, we know what sort of correlations we are looking for and we are finding them.

Our question remains: What may we conclude if these criteria are met, i.e., if we find that every type of mental process does indeed have its neural counterpart which mirrors every change and nuance of its correlated mental process? We may suppose that each type and degree of, e.g., pain, is invariably correlated with a specific type and degree of neural activity. We have, let us say, a perfect isomorphism of all mental processes with specific neural processes. Further, we are able to account for these mental processes, e.g., sensation, memory, etc., in terms of neural processes. Would this mean that, e.g., sensations are really neural processes?

It would surely be folly to ignore the findings of neurophysiology. In point of fact, it does look more and more as if there is a neural counterpart to every mental process. For example, physiologically inclined psychologists seem on the verge of discovering the mechanisms of memory. (We are excluding mental processes which may be adequately explained in terms of behavioral dispositions.) But the problem remains of how to *interpret* these empirical findings.

---

[35] *Ibid.*, p. 101.
[36] *Ibid.*
[37] *Ibid.*, p. 102.
[38] *Ibid.*, p. 103.

There is the Identity theorist's interpretation which identifies the mental processes with the neural ones. Place would say that "Consciousness is a brain process and *nothing else*" [39] and Smart would say either that "Sensations are nothing over and above brain processes" or that what sensations really are, what their true nature as revealed by science is, is neural processes.[40]

There is, as I have mentioned, a kernel of truth in this contention. It is this: pains and other sensations undoubtedly do have physiological (especially neural) components. They do not exist, after all, in a vacuum. So far as we could ever tell, they occur only in sentient organisms. They occur in living bodies (e.g. persons) and it is difficult to see how anyone could produce any evidence for the view that they occur in some non-located, indivisible, Cartesian mind-stuff which may or may not be inexplicably causally related to our bodies. The findings of neurophysiology do have a direct bearing on the mind-brain problem. If there were no detectable correlations between experiences and certain brain processes, some version of Cartesianism might have to be adopted. But few, if any, Identity theorists have commented on the assumptions, findings and conclusions of the scientists who are actually conducting empirical research into the working of the brain. If the Identity Theory is a genuine scientific hypothesis, surely, one would think, it has not escaped the notice of psychologists, physiologists and neurophysiologists. Materialistic interpretations of mental processes have been available, albeit in a crude form, since the time of Hobbes, and even as far back as Democritus. There is nothing logically new in its modern versions, though present day materialists can appeal to well-established mind-brain correlations which were unavailable to Democritus and Hobbes.

As we saw earlier, some neurophysiologists see a gap between the mind and the brain that seems unbridgeable. A survey of the neurophysiological literature reveals that the majority of those scientists conducting experiments on the brain assume some form of dualism. There are not many who would agree with Lashley that "No activity of mind is ever conscious".[41] This is an indication that the Identity Theory and Dualism are *interpretations* of facts and not independently testable theories about them.

[39] Cf. pp. 102-03 of "Is Consciousness a Brain Process?".
[40] Cf. pp. 163 and 165 in "Sensations and Brain Processes".
[41] K.S. Lashley, *The Neurophysiology of Lashley,* ed. by F.A. Beach *et al.,* p. 532.

That the majority of researchers in a field believe in (or assume) some sort of dualism does not, *ipso facto,* mean that it is true. It does mean, however, that we should not rashly reject the assumption of some sort of dualism before making every effort to see whether something can be said in its favor.

Most neurophysiologists, as has been stated, assume some form of dualism and causal interaction between mental and neural processes. However, they usually talk in terms of physical-to-mental causation or dependence and rarely in terms of mental-to-physical or mental-to-mental causation. If they assume any philosophical position, it is usually an epiphenomenalistic position—the view that mental processes are entirely dependent on physical ones and are themselves inefficacious—but they do not embrace materialism. It may be instructive to see why they assume some sort of dualism.

As we saw earlier,[42] Penfield talks in terms of an impasse between mental and physical processes which somehow eludes our grasp. He adds the following:

If he [the neurophysiologist] should state that nerve impulses moving in certain patterns are one and the same thing as mind [i.e. state the Identity Theory], he accomplishes little for his future work except to deprive himself of a useful working terminology.... However it is expressed, he must think either of a parallelism or a back and forth relationship.[43]

Here Penfield says, in effect, that adoption of the Identity Theory would hamper rather than aid research. It would, he says, deprive the scientist of a useful way of referring to the phenomena under investigation. Penfield has done extensive work in brain-mapping, i.e., determining which parts of the brain are responsible for certain behavior (e.g. motor responses) and certain experiences. Let us see why he finds a dualistic terminology useful or perhaps even finds it impossible to proceed without it.

The following is a sample of the way in which Penfield (in common with most other neurophysiologists) describes some of his experiments concerned with correlating neural and mental processes in "The Interpretive Cortex".

There is an area of the surface of the human brain where local electrical stimulation can call back a sequence of past experience....The sights and sounds, and the thoughts, of a former day pass through the man's mind again.[44]

[42] Above, p. 26.
[43] Penfield, *Speech,* p. 8.
[44] In *Science,* p. 1719.

Occasionally....gentle electrical stimulation in this temporal area, right or left, has caused the conscious patient to be aware of some previous experience. The experience comes back to him in great detail.[45]

The action of the living brain depends upon the movement, within it, of "transient electrical potentials traveling the fibers of the nervous system". This was Sherrington's phrase. Within the vast circuits of this master organ, potentials travel, here and there and yonder, like meteors that streak across the sky at night and line the firmament with trails of light. When the meteors pass, the paths of luminescence still glow a little while, then fade and are gone. The changing patterns of these paths of passing energy make possible the changing content of the mind. The patterns are never quite the same, and so it is with the content of the mind.[46]

This electrode....produces no more than elementary sight when applied to visual cortex. The patient reports colors, lights, and shadows that move and take on crude outlines. The same electrode, applied to auditory cortex, causes him to hear a ringing or hissing or thumping sound....Thus, sensation is produced by the passage inward of electric potentials.[47]

[When an electrode is] applied to the various sensory areas of the cortex, it causes him to have crude sensations of sight or sound or body feeling.[48]

The neurophysiologist stimulates the brain electrically or chemically and the subject then describes what he sees, thinks of, remembers or feels. It seems difficult to believe that there are not two different sorts of phenomena or at least two different aspects of one process occurring here but only one sort, i.e., physical. The brain is stimulated, thus producing certain neural processes in the subject and, simultaneously, the subject undergoes a certain experience. The scientist can observe, or at least measure, the neural processes but only the subject himself observes, or could observe, the experiences. Here "observe" means "have", which means "is aware of", in the sense that the subject need not and could not *infer* what these experiences are but is non-inferentially aware of them. Strictly speaking, a neurophysiologist cannot actually observe the subject's neural processes. Using an electroencephalogram, he can record the electrical activity of some of the subject's neural processes. And of course he can observe the subject's brain. But he must rely on the subject's testimony if he is to learn of the existence and nature of the latter's experiences. The scientist may be able one day to infer accurately what experiences a subject is having. But the subject himself could never infer this—he *has* the experiences.

[45] *Ibid.*, pp. 1719-20.
[46] *Ibid.*, pp. 1721-22.
[47] *Ibid.*, p. 1722.
[48] *Ibid.*, p. 1724.

In all the quotations cited above, we see the difference between the physical stimulation of the brain and the consequent neural processes on the one hand, and on the other hand the mental processes which the subject reports. This is why neurophysiologists speak in terms of a dualism of physical and mental processes and of *correlating* them. The things observed by the scientist and those observed or experienced by the subject seem so radically different that it strikes one as an incredible suggestion that there is really only one type of process occurring, viz., the neural ones.

In the case of pain, it would seem that the physical processes, including the neural ones, either cause, or are manifested to the subject in the form of, the mental process, i.e., his experience of pain. (There may be no way to choose which is the case by *observation*—it may be a matter of our interpretation of the facts.)

Penfield sees as clearly as any Identity theorist that the relationship between mental and neural processes is a peculiarly intimate one. For example, in the third quotation above, he says that the changing patterns of cerebral processes "make possible" the changing content of the mind. He adds that the contents of the mind mirror the patterns of the brain. Penfield seems to think that some sort of dualism is forced upon us since we are presumably making mental-physical *correlations*. One could not correlate something with itself, therefore mental processes, it may be argued, cannot be *simply* neural processes, though they may indeed be a manifestation of neural processes.

The Identity theorist could argue that the experience of pain is itself nothing but those neural processes which the neurophysiologist describes as occurring *simultaneously* with the experience. "What *else* could the having of pain consist of than the neural processes?" the Identity theorist can ask. "The dualist", says Smart, "cannot really say that an experience can be composed of nothing. For he holds that experiences are something over and above material processes, that is, that they are a sort of ghost stuff. (Or perhaps ripples in an underlying ghost stuff)".[49]

We shall see in the next chapter how something can be *composed* of elements, yet not be "identical with", or "nothing but", or "really" those elements. As to Smart's reference to a "ghost stuff", traditional dualism has always claimed that mental processes are not *composed* of anything (physical).

[49] Smart, "Sensations", p. 170.

It is my contention that neither the materialists nor the dualists characterize these mental processes satisfactorily. Materialists often sound as though they are trying to reduce mental processes to the extent that there is nothing 'mental' left of them. Instead of explaining them, they try to explain them away. Dualists, on the other hand, usually seem to regard the mind as a sort of incorporeal thing or substance which in an unknown and seemingly incomprehensible way, interacts with the brain and thus the body. In the sequel, I shall be arguing for the view that pains and other mental processes have a neural component but they are neither reducible to, nor are they "really", neural processes. I shall also argue for an *epistemological* dualism which is manifested in an asymmetry in the way I know about my sensations and the way another does.

While an experimenter could in principle observe the neural aspect of pain, neither he nor anyone else could *observe* the sensation or feeling aspect of pain. Nor could he be in the same epistemological relationship to the subject's pain, unless he himself were the subject of his own experiment. The fact that only the subject is, and could be, aware of the pain in the way in which he is, is one of the facts which is vital to the concept which we have of pain and other sensations. (Of this, more later.)

Place draws an analogy between the identification of lightning with the motion of electric charges and the identification of sensation with brain processes:

Thus we conclude that lightning is nothing more than a motion of electric charges, because we know that a motion of electric charges through the atmosphere, such as occurs when lightning is reported, gives rise to the type of visual stimulation which would lead an observer to report a flash of lightning.[50]

He then adds, as we saw before, that "in order to establish the identity of consciousness and certain processes in the brain, it would be necessary to show that the introspective observations reported by the subject can be accounted for in terms of processes which are known to have occurred in his brain".[51]

A cloud consists only of a mass of tiny particles in suspension. What we call a cloud, when seen from a distance, may be called by a different name—a mist or fog—when we are immersed in it. For example, when approaching a mountain we may see clouds enveloping its summit. Once we have reached the summit, however, we would say that we are

---

[50] Place, "Consciousness", p. 106.
[51] *Ibid.*

enveloped in a fog or mist. And, while we are so enveloped, we can observe the mass of tiny particles which, when seen at a distance, gave rise to our observation of a cloud.

In the same way, Place implies, what we observe under certain conditions as an instance of consciousness is seen, when observed under different conditions, as a process in the brain. Just as a cloud can be completely explained in terms of, and therefore identified with, a mass of tiny particles in suspension, so consciousness may be completely explained in terms of, and therefore identified with, a process in the brain.

In the case of clouds and masses of tiny particles in suspension, nothing more than ordinary processes of observation are required. However, as Place says, "The operations required to verify statements about consciousness and statements about brain processes are fundamentally different".[52]

He then cites the example of lightning and a motion of electric charges. Lightning is observed by "ordinary processes" but the motion of electric charges is determined by "special scientific procedures",[53] yet the former is identified with the latter. The lightning and the electric charges are the same event, viewed in two entirely different ways. So, he reasons, consciousness and a brain process may be the same event, viewed in radically different ways.

But how do we decide whether consciousness and brain processes are the same event observed in two radically different ways or are different events? We may agree with the Identity theorists that all cases of consciousness and sensation have a neural counterpart, but how are we to decide whether or not to identify the mental and the physical process?

As we have seen, Place suggests two criteria of identity, one concerning explanatory power and the other concerning identity of spatiotemporal reference.

However, neither of these criteria seems to have been met. As we have seen, neurophysiologists have not regarded their discovery of an invariable correlation and strict isomorphism between consciousness and brain processes as an adequate explanation of consciousness. They reason that the brain process must somehow give rise to the sensation or consciousness but are unable to conjecture how this (apparently) causal process comes about.

[52] *Ibid.*, p. 105.
[53] *Ibid.*, p. 106.

The discovery of a neural correlate for consciousness has not been seen as confirming the Identity Theory. True, some of the mystery of the mind will have been removed if it is firmly established that every mental event has a neural correlate. However, so long as the mental and physical process are interpreted as two utterly dissimilar processes, the neural process will never suffice as an explanation of the mental process. For, it will be asked, how does the neural process bring about the mental process? So long as we think, in Cartesian terms, of mental processes as a type of process in one type of medium or substance—unextended and indivisible—and of physical processes as occurring in a completely dissimilar type of substance—extended and divisible—just so long will we regard the gap between the mental and the physical as impassable. When seen in their true light, these mental processes are not different processes from the neural ones, but neither are they merely identical with them.

The second criterion of identity, that the two sets of observation must refer to the same point in time and space, has also not been regarded as fulfilled. It has not been regarded as merely unfulfilled in fact—it has been deemed by some thinkers impossible of realization. This is because certain mental processes, if not all of them, have been considered to be necessarily non-spatial.

Hume, for instance, says the following:

For can any one conceive a passion of a yard in length, a foot in breadth, and an inch in thickness? Thought therefore and extension are qualities wholly incompatible, and never can incorporate together into one subject....What is extended must have a particular figure, as square, round, triangular; none of which will agree to a desire....nor can a smell or sound be either of a circular or a square figure.[54]

If experiences cannot be located in space, then they cannot be identical with brain processes. A contemporary philosopher who echoes Hume's sentiments is Jerome Shaffer. He argues that it makes no sense at all to talk about experiences and other mental processes as being located somewhere in the body. He then claims that since it makes no sense to speak of experiences as in the brain, it also is senseless to consider them as not in the brain.

The fact of the matter is that we have no rules in our language either for asserting that C-states [55] have a particular location or for denying that they have a particular

---

[54] David Hume, *A Treatise of Human Nature*, pp. 213-14.

[55] Shaffer defines "C-states" as "mental states, e.g., feeling pain, having an after-

location. So we have here a case in which it is senseless to apply the criterion of same location. But the Identity Theory still will not do, because if it is senseless to apply one of the criteria for identity then it is also senseless to claim that there is identity.[55]

The issue of locatability of experiences is discussed at length both here and on 99 ff. Here we should note that it is odd to speak of "observing" an experience at all, let alone observing it in a particular place. What could "observing" an experience mean other than *having* or *undergoing* the experience? We can specify how long a pain lasts but how would we go about specifying where we had the experience of pain? It makes no sense to say that another person observes my experience at some place in my body and it makes no sense to say that I observe it anywhere. As I argue in the sequel, experiences are not the sort of thing which anyone could observe. Nor could we observe concepts or fictional characters in space.

Thus one of the criteria of identity which Place specifies is incapable of fulfillment. This means that it will never be possible empirically to confirm the identity of, e.g., pains and brain processes. What we can do is determine which brain processes are correlated with certain experiences by examining how long an experience is undergone and detecting the specific brain processes which occur during that time. We may be able to determine the location of the brain processes. We know what it would mean to locate them. But we are unable to determine the bodily location of the experience correlated with them—we do not know what it would *mean* to locate it.

An Identity theorist could argue that if sensations are brain processes, then they must have all the properties of these brain processes. Smart, for example, defends this view:

All that I am saying is that "experience" and "brain process" may in fact refer to the same thing, and if so we may easily adopt a convention (which is not a change in our present rules for the use of experience words but an addition to them) whereby it would make sense to talk of an experience in terms appropriate to physical processes.[57]

As we argue in the sequel, it seems to be impossible to make any sense of locutions which involve locating experiences in space or, what

---

image, thinking about a problem, considering some proposition, etc.", "Could Mental States be Brain Processes?", p. 813.

[55] *Ibid.*, p. 816.

[57] Smart, "Sensations", p. 169.

is the same thing, assigning spatial characteristics to them. Smart is mistaken in thinking that talk about locatable experiences would involve no changes but only additions to the current rules governing the use of experience-words. If what is senseless by the present rules becomes sensible under new rules, then conceptual revision of a serious nature will have occurred.

There is nothing wrong with linguistic or conceptual revision *per se*. On the contrary, such revision can signify a magnificent achievement, for example, Einstein's revision of the concepts of space and time. But the case of pain is utterly different. Identity theorists seem to think that the only way in which Cartesian dualism can be revised is to substitute a materialistic monism for it. It is my view that these are *not* the only two alternatives. I too wish to revise the notion of mentality found in Cartesian dualism but I shall try to show that we can correctly characterize pain and account for its mentality *without* resorting to a reductionist materialism.

Smart sees no harm in adopting a convention which would enable us to say that one of our experiences occurred three inches behind our left eye and another occurred just one eighth of an inch below the first experience. However, assuming that it makes sense at all, it is dubious how useful it would be to adopt a convention which enabled us to assign a bodily location to experiences. The reason for this is that it is difficult to see when the question of the bodily location of an experience would *arise*.

When we ask someone where he had a certain experience, he replies in terms of where *he* was, for example, "In the basement" or "In Rome". It is far from apparent whether we could find a *use* for a question about our experience which required a response such as "Three inches behind the right eye". In short, recommendations for revising the conventions of ordinary language must be viewed with a critical eye, particularly when there seems to be no purpose whatever in adopting the recommended changes. (The last chapter of this book deals at length with recommendations about language.)

J.J.C. Smart has, at times, argued as if the Identity Theory were a scientific thesis. He sees the adoption of the Identity Theory as one of the final steps on the way to the time when all events will be explicable solely in terms of physical processes. We shall now see whether Smart makes good his claim that science could reveal to us the true nature of our sensations and experiences.

The Identity Theory has been expanded and defended in several papers and a book by Smart. He propounds a thoroughgoing materialism, claiming that persons are "simply very complicated physico-chemical mechanisms" [58] and that the picture of the world which sub-atomic physics gives us is "truer" than that of common sense.[59]

Smart holds that "it may be the true nature of our inner [sic] experiences, as revealed by science, to be brain processes, just as to be a motion of electric charges is the true nature of lightning, what lightning really is" [60] but that "there is no conceivable experiment which could decide between materialism and epiphenomenalism".[61]

If we could conduct no experiment which would enable us to decide between the Identity Theory and some other theory, then we cannot use scientific means to "reveal" the "true nature" of our "inner experiences". Perhaps Smart means that the physical-mental correlations established by science do not conflict with, or that they lend plausibility to, the speculation that experiences are nothing but brain processes. Whether or not he means this, assuming that we actually are unaware of the nature of experiences, e.g., pains—that we do not really know *what they are*— then presumably some sort of analysis and/or interpretation is required if we are ever to be set straight about them.

If there are no scientific procedures which could one day reveal to us the "true nature" of, e.g., our pains, then perhaps we can decide the issue between Identity and Dualism on non-empirical grounds. In fact, Smart argues that since we cannot decide empirically between materialism and epiphenomenalism, we must opt for the former view because the latter theory is of the type which "offends against the principles of parsimony and simplicity".[62]

In the following chapter we shall be considering the conceptual issues raised by Smart. Here we shall examine some of his claims about the empirical aspects of the mind-brain problem. In Smart's first paper defending the Identity Theory, he says:

[58] Smart, *Scientific Realism*, p. 2.

[59] *Ibid.*, p. 47.

[60] *Ibid.*, p. 93. This is a repetition of his argument on pp. 164-65 of "Sensations and Brain Processes" that "modern physical science" tells us what "the true nature of lightning" is. He concludes that "what lightning really is, what its true nature as revealed by science is, is an electrical discharge".

[61] Smart, "Sensations", pp. 171-72.

[62] *Ibid.*, p. 172.

There does seem to be, so far as science is concerned, nothing in the world but increasingly complex arrangements of physical constituents....That everything should be explicable in terms of physics (together of course with descriptions of the ways in which the parts are put together....) except the occurrence of sensations seems to me to be frankly unbelievable.[63]

Smart suggests here that science alone is fit to determine what are the constituents and features of the world. Few scientists would make such a claim. Most realize, as Ryle reminds us, that the formulae of scientific theories "are constitutionally speechless about certain sorts of matters [e.g., chairs, tables, colors, experiences], just because they are *ex officio* explicit about other, but connected matters [ultimate particles]." [64] Furthermore, the picture of the world which sub-atomic physics gives in terms of elementary particles is not at all incompatible with, but is complementary to, our common sense picture of the world in terms of chairs and tables. Wittgenstein aptly says of scientists who claim, e.g., that the floor on which we stand is not really solid (as common sense deems it to be) that the picture of the atomic structure of matter "was meant to *explain* the very phenomenon of solidity".[65] This issue arises again in the case of sensations and brain processes.[66]

Smart's suggestion that everything is, or will be, explicable in terms of physics is incredible. As Ryle says in *The Concept of Mind*: "Physicists may one day have found the answers to all physical questions, but not all questions are physical questions".[67] Just to give an example, the question why Smart holds the Identity Theory is not a physical question, or at least not solely physical. All issues involving purposes, intentions, motives, and reasons, not to mention historical and moral questions, are inexplicable solely, if at all, in terms of physics. Smart anticipates the time when "even the behavior of man himself will one day be explicable in mechanistic terms",[68] not seeming to realize that an explanation of behavior in mechanistic terms need not exclude explanations in other terms, e.g., psychological and anthropological. (Adequate discussion of this issue would involve much more space than could be devoted to it here.)

Smart's version of the Identity Theory is virtually the same as that of Place's, except, as we have seen, that Smart explicitly says that the

[63] *Ibid.*, p. 161.
[64] Gilbert Ryle, *Dilemmas*, p. 78.
[65] Ludwig Wittgenstein, *The Blue and Brown Books*, p. 45, emphasis mine.
[66] Cf. below, pp. 72 ff.
[67] *Op. cit.*, p. 76.
[68] Smart, "Sensations". p. 161.

issue between the Identity Theory and Dualism, (e.g. Epiphenomenalism) is not empirical but logical. He says, at times, that the issue is purely empirical only in the sense that we can test, and have of course already tested, to determine whether it is the heart or the liver or the brain which is involved with sensations (and other mental processes, J.O.) but otherwise simplicity and plausibility are the appropriate criteria.[69]

Smart says that "in so far as a sensation statement is a report of something, that something is in fact a brain process. Sensations are nothing over and above brain processes.[70] He is not, he says, claiming that sensation statements are translatable without loss into brain process statements nor is he claiming that the logic of these statements is the same. Both he and Place are claiming that there is a contingent identity between sensations and brain processes. Sensations are *in fact,* not of necessity or by definition, brain processes.

In *Philosophy and Scientific Realism,* Smart discusses the sensation of pain.

....suppose that I report that I have a pain....There seems to be some element of 'pure inner experience' which is being reported, to which only I have direct access. You can observe my behavior, but only I can be aware of my own....pain....The notion of pain seems essentially to involve the notion of distress, and distress is perhaps capable of an elucidation in terms of a characteristic behavior pattern. But this is not all that a pain is: there is an *immediately felt sensation* which we do not have in other cases of distress.[71]

He adds that in the case of having an after-image, there is "not this 'emotional' component of distress",[72] so he concentrates on reports of having after-images rather than those of having pains. This is most unfortunate. It is the bodily feeling of distress or anguish, the "immediately felt sensation" and its relationship to neural processes, which is the point at issue where pain is concerned. We need have no *feelings* where after-images are concerned.

Smart asserts that "the goings on which are reported [e.g. that which constitutes the feeling of pain] are in fact brain processes".[73] Sensations are not non-physical processes, he says, they are processes occurring in the brain. When we report a pain to the dentist, we are reporting not a

---

[69] *Ibid.,* pp. 171-72.
[70] *Ibid.,* p. 163.
[71] *Op. cit.,* pp. 89-90, emphasis mine.
[72] *Ibid.,* p. 90.
[73] *Ibid.,* p. 91.

mental (non-physical) process but a purely physical one. When one says, "I have a pain", this is "to the effect of 'what is going on in me is like what goes in me when a pin is stuck into me' ".[74] Thus the experience of having a pain is to be classified as like the experience one has when one is, e.g., stuck with a pin or burnt by a match. Smart says that "this likeness, on my view, must consist in a similarity of neurophysiological pattern".[75]

The plain man would say that what goes on in him when he is stuck with a pin or burnt by a match is, to say the least, a feeling of pain. Smart claims that this "immediately felt sensation" is, to say the most, a particular pattern of neural processes. The plain man would say that incidents of being stuck with pins and being burnt by matches are alike in that they all involve a feeling of pain. Smart points to the similarity of neurophysiological patterns.

Thus we may well be mistaken in thinking that our feelings of pain are, or at least involve, mental processes or aspects, according to Smart. There are, he says, no inner aspects or qualities (of neural processes) which are distinct from the (theoretically) publicly observable neural processes.[76] (Here he repudiates the dual aspect view.) There are only neural processes which, when they occur in particular patterns, cause us to report them as pains.

He claims that there is a 'strict' identity between sensations and brain processes. This is the sort of identity holding between the Evening Star and the Morning Star, or better still, he says, between lightning and an electrical discharge. Though the Evening Star and the Morning Star are one and the same thing, neither could informatively be said to be revealed as the *true nature* of the other, as science has revealed what lightning *really is,* viz., electrical discharge. It would be odd to say that one star is "composed" of another. Smart's theory is that sensations are *really* neural processes, because they are *composed* of nothing but neural processes.

But to say that lightning is really electrical discharge does not mean that there is no such thing as lightning. It would be absurd to say that "A = B" and then to add "And there is no A". Yet this seems to be the implication of what Smart is saying about, e.g., pains. To say that they are really nothing but neural processes appears to mean that they could

[74] *Ibid.,* p. 96.
[75] *Ibid.,* p. 95.
[76] *Ibid.,* p. 94.

not also be "immediately felt sensations".[77] It would seem that even if there is a sensation-brain process identity, it cannot be a 'strict' identity as that which holds between the Evening Star and the Morning Star or between lightning and electrical discharge.

If Smart is denying that we experience sensations of pain then his view is not worth serious attention. This issue is similar to the one discussed on pages 71 ff. The theory that material objects, e.g., tables and chairs, are composed of a vast number of elementary particles in no way means that material objects are not *real*. On the contrary, the atomic theory was adduced in order to *account* for material objects. Here $A = B$ (material objects = elementary particles) and we see clearly that it would be absurd to say that "$A = B$" ("Pains = neural processes") and then conclude from this that there are no A's. It is evident that to purport to account for phenomena, e.g., material objects or pains, and yet to deny their existence is to contradict oneself.[78]

We must assume, then, that what Identity theorists are saying is that people have experiences such as being in pain but that their being in pain involves nothing but the occurrence of certain brain processes, e.g., the firing of certain neurons. If Identity theorists themselves are unable to envision experiments or situations such that their conjecture will be rendered testable, perhaps we can do so.

It is theoretically possible that some day neurophysiologists will be able to observe all one's physiological and neural processes as one undergoes experiences. Then, perhaps, we will be able to confirm or disconfirm the Identity Theory. If sensations are nothing really but brain processes, it would be, in principle, possible for us to observe them in some way. If they are physical, it must be theoretically possible to observe them intersubjectively.

We shall assume that neurophysiology has developed an instrument whereby we can detect every feature of every physical (including neural) process occurring in a subject's body. A subject is connected to this machine, and in order to produce the desired experience, we pour cold

---

[77] Both A.J. Ayer and Jenny Teichmann have voiced suspicions that this is what the Identity Theory comes to. Cf. Ayer's review of D.M. Armstrong's *A Materialist Theory of the Mind,* called "Mind and Matter", in the *New Statesman* and see Teichmann's "The Contingent Identity of Minds and Brains".

[78] One Identity theorist, D.M. Armstrong, really does seem to explicitly deny that there are sensations in the ordinary sense of the term. He interprets the sensation of pain as a perception which in turn is analyzed solely in terms of the acquisition of beliefs, potential or actual. See *A Materialist Theory of the Mind.*

water on an exposed nerve in his mouth. As Feigl says, "the crux of the mind-body problem consists in the interpretation of the relation between raw feels and the neural processes".[79] It is this relation which we are now considering.

In the case at hand, we pour cold water on the exposed nerve and the subject has an excruciating pain. Does his raw feel, the feeling of pain, consist of nothing but neural processes or does it consist of something over and above or entirely different from them? Do we have here an identity or a correlation? I shall try to show that there is a correlation here and not an identity.

Let us call the machine which reveals all of one's physical processes an "S-ray" machine where "S" stands for "sensations". Neurophysiologists have already found that pains show up on an electroencephalograph (EEG) in the form of a spike pattern. The needle registering the electrical activity of the brain "jumps" when the subject experiences pain and this leaves a spike on the EEG. We can assume that our S-ray machine shows that one's C-fibers (the neural processes associated with, or reputedly identical with, pain) themselves do indeed fire in the appropriate way to produce a spike pattern on an EEG. The S-ray machine reveals everything physical that happens in the person from the time we pour the water onto his exposed nerve to the time when he groans, clutches at his jaw and finally sighs with relief when we cease to apply the water.

Here, then, is a case where nothing (physical) is hidden from us. We could imagine the S-ray machine to be a sort of super x-ray machine or T.V. camera which shows us everything that is going on in the subject's body as he stands in front of the apparatus. Now that we can observe every physical (including neural) process in him, is there anything which we still cannot observe? If the Identity Theory were true, then surely the answer to this question would be "No". If one's being in pain (or the experience of pain, or feeling of pain, or sensation of pain) is nothing but a neural process occurring in one's body, then surely if all one's physical processes are exposed to view, one's experiences must be revealed at the same time. (A 'strict' identity is a strict *identity*.) We should even be able to take a picture of someone's sensations, if we could take one of his neural processes.

---

[79] Herbert Feigl, "The 'Mental' and the 'Physical'" in *Minnesota Studies in the Philosophy of Science*, ed. by H. Feigl, M. Scriven and G. Maxwell, p. 446.

If there is an objection to the effect that in the experiment as outlined above, the experimenter still cannot "directly" observe the subject's neural processes because, e.g., an instrument is involved, then we can easily imagine that we have managed to breed people who are in a way transparent, except that all their physical processes can be observed. Would we now say that we could observe their sensations? Have we finally rendered the Identity Theory testable?

In one sense of "observe" and in one sense of "sensation", I think that we might all agree that, were all our neural processes visible, we could observe each other's sensations. But this does not involve the ordinary use of the term "sensation".

"Observe" means something different in each of the following two sentences:

1) I observed the castle for two hours.
2) I observed my pain very closely, as the doctor had asked me to do, and described it as "gnawing" rather than "stabbing".

In the first sentence "observe" means to watch or to see and it involves the use of one's eyes. In the second sentence "observe" means to pay close heed to or to attend to, and it does not involve the use of any such sense organ.

According to my dictionary, "sensation" is commonly used to refer to the subjective element or physical feeling consequent on and related to some portion of the bodily organism. Corresponding to this usage would be the common usage of "pain" to refer to the sensation which one feels when hurt. This is what is referred to by the ordinary use of "sensation". It is what one *avoids* except where it is required to bring about a less painful situation, e.g., going to the dentist in order to get a cavity filled.

But "sensation" has another use. It is sometimes found in physiological literature. In this second usage, one could have a sensation and not be aware of it. For example, a physiologist may speak of a taste cell sending a taste sensation to the brain. Or he may speak of eliminating pain sensations from consciousness. There is the suggestion in these cases, particularly in the first case, that sensations can be potential as well as actual and that they could exist, for a while at least, unfelt. What is actually referred to here, I think, is the neural aspect of sensation.

With these distinctions in mind, let us return to the case of the "transparent" subject, all of whose neural processes are observable. Would we be able to observe his sensations? Not in the relevant sense. We could never see his feeling of pain because it is not observable in that sense to anyone at any time. That is, we could never see or watch his pain and neither could he, though he obviously is aware of his pain. In the first sense of "observe", neither the experimenter nor the subject could observe his raw feel. But there is another sense of "observe", i.e., pay heed to or attend to. Could we observe his pain in this sense?

The answer to this question is "No". Any pain one feels is *ipso facto* his own pain. Even if someone invariably felt pain whenever his friend did, or could somehow transfer pain from his friend to himself, the pain he felt would be *his* pain. The concepts and the criteria of identity of persons and of pains which we have just do not allow us to count one person's pain as that of another. It is impossible even to see what it could mean to have another's pain. Incidentally, it is also difficult to imagine circumstances where it would be instructive to say "I can have only my own pain". The statement that one can have only one's own pain, and thus be aware of it in *that* sense, is a remark about our concepts and criteria, not an empirical claim or discovery (except in so far as concepts are functions of empirical facts).

I am not suggesting that we need always be in doubt about the pains of others. One need not *have* (and could not have) another's pain in order to know that he is in pain. If a sceptic insisted that we really do not know that our subject (who is writhing and groaning after having the water treatment on his exposed nerve) is in pain, it would be perfectly in order to remind him that we know pain when we *see* it. But this is not to claim that we can see his feeling of pain. We mean we see *that* he is in pain and to doubt this would mean that we do not know what pain is.

I have argued that pains are logically nontransferable from one person to another. As Terry Forrest shows in "P-Predicates", if this is true of mental processes, it is true also of physical processes of persons, e.g., one could not have another's cut or bruise.[80] After a heart transplantation, one can have the heart of another but then the heart *is* no longer the other's. The same, I submit, would apply to pains if we could somehow transplant pains.

If the Identity Theory were true, we should be able literally to see another's pains (as we see his cuts or bruises) if we could see his neural processes. But this, I have argued, we cannot do. Here is at least one feature of pains which shows that they are not simply neural processes.

---

[80] In *Epistemology*, ed. by Avrum Stroll, p. 101.

Were people transparent in the sense explained above, we might never doubt whether anyone were in pain, etc., because then we could actually see if his C-fibers were firing in the appropriate way or not. But, as we have seen, we would still not be in the least inclined to say that we are aware of his pain in the same way in which he is aware of it, i.e., in the sense that he HAS the pain. This would be the case only where the experimenter is also the subject.

Wittgenstein, I think, was the first one to envisage such an experiment. In *The Blue Book,* he says the following:

Let us imagine such an experiment crudely. It consists in looking at the brain while the subject thinks. And now you may think that the reason why my explanation is going to go wrong is that of course the experimenter gets the thoughts of the subject only *indirectly* by being told them, the subject *expressing* them in some way or other. But I will remove this difficulty by assuming that the subject is at the same time the experimenter, who is looking at his own brain, say by means of a mirror. (The crudity of this description in no way reduces the force of the argument.)

Then I ask you, is the subject-experimenter observing one thing or two things?.... The subject-experimenter is observing a correlation of two phenomena. One of them he, perhaps, calls the *thought.* This may consist of a train of images, organic sensations....The other experience is one of seeing his brain work. Both these phenomena could correctly be called "expressions of thought"; and the question "where is the thought itself?" had better, in order to prevent confusion, be rejected as nonsensical.[81]

These remarks can also be applied to the case of sensations. On the one hand the subject sees his neural processes and on the other he feels the pain. Is he observing one thing or two things? If there is only one thing being observed and it is physical, then the Identity Theory holds. If there is a correlation of two phenomena, however, it will not do to say that pains are nothing but neural processes. They are "inner" processes in the sense that they are felt and not seen.

The neural processes which the subject sees are the neural aspect of his sensation and his experience is the mental aspect of his sensation. This is why it would be confusing if someone were to ask "But what is the sensation itself?"—all that we can do is to indicate the two different aspects of sensation and remind the questioner that the ordinary use of "sensation" indicates the experiential aspect of sensation since this is the aspect of which one is normally aware.

[81] *Op. cit.,* pp. 7-8.

Though the subject never has the "raw feel" without having the appropriate neural processes, and never has the latter without having the former, we cannot say that the raw feel is *nothing but* the neural process. Were we to say this, then it would make no sense to speak of raw feels being *correlated* with neural processes. If the Identity Theory were true, then no sense could be made of the *correlations* which neurophysiologists are discovering. The fact that sensations and neural processes can be identified independently of one another, by the subject and neurophysiologist respectively, also indicates that there cannot be a 'strict' identity here.

CHAPTER THREE

CORRELATION, IDENTITY AND SUBSTANCE—
SOME CONCEPTUAL ISSUES

I have stated the mind-brain problem as follows: What is the relation between mental processes, e.g., sensations, and the neural processes which occur invariably and simultaneously with them? Materialistic Identity theorists claim that what are called "sensations" are not different from neural processes but that, for some (usually unstated) reason, the neural processes can be referred to in two different terminologies, viz., that of physical science and that of subjective experience. R.J. Hirst, who holds what he calls an Identity Hypothesis, explains the existence of a dualistic terminology by reference to the radically different mode of access which an observer of someone's 'experience' (neural processes) has from that of the person undergoing the experience.[1] The pain which I feel (or my feeling of pain) is one and the same thing, says Hirst, as the neural process which you could (in principle) see. However, I experience or observe it very differently from the way in which you do because my mode of access to the experience is entirely different. In other words, my relationship to my neural processes (experience of pain) is utterly different in character from your relationship (mode of access) to them. Hence both the subjective and the objective modes of speech to refer to these processes are necessary and complementary (not incompatible).

Surely Hirst is right when he says that an observer and one in pain have utterly different modes of access to the pain. But it is misleading to say that they are observing "the same thing". He says that "owing to the radically different modes of access, that event [the feeling of pain] appears as a pattern of behavior or of brain activity".[2] As I argued

---

[1] Presumably U.T. Place, who also stresses the radically different ways of observing "the same phenomenon", would hold such a view on the need for a dualistic terminology.

[2] *The Problems of Perception,* p. 193.

earlier,[3] I cannot see (observe) your feeling of pain and neither can you though both of us could in principle see your neural processes. This indicates that what the subject feels and what another sees cannot be simply identical, though of course what another sees (i.e. the neural processes) may be a feature of what the subject is undergoing.

But most Identity theorists neglect the asymmetry between the way in which I am aware of my pain and the way in which others might be aware of it. Being materialists, they are committed to denying the reality of processes which in principle are available only to one person. However, as John Wisdom rightly says, "The peculiarity of the soul [mind] is not that it is visible to none but that it is visible only to one".[4] Thus the peculiarity of the experiential aspect of pains is not that it is undetectable by any of the five senses or by any mechanical instruments which we could devise but that each person can feel or be "aware" of (in the sense of "have") only his own and never anyone else's. In this sense, consciousness is indivisible. This is something Descartes insisted upon, though he conceived of the mind as a thinking substance and did not consider the possibility of mental processes occurring without a mental substratum. This seems to be the conceptual model of Identity theorists also, as we shall see.

If we insist that there are sensations apart from, or over and above, the neural processes, then we must try to elucidate the relation between these two phenomena (or aspects of one phenomenon). If we say that the two events are merely parallel, that they merely happen to occur together, then we ignore the findings of the neurophysiologists that sensations can be produced whenever certain neural processes are produced and that the former invariably accompany certain sorts of neural processes, i.e., we ignore the fact that there is *some* sort of relationship or connection between the two phenomena.

If the relationship is one of 'strict' identity, as Smart claims, then the statement "Sensations are brain processes" is true. But if this statement is true, then so is its converse: "Some brain processes are sensations". The problem of accounting for sensations by the techniques of science has, therefore, not been solved. Now some brain processes, namely, those which constitute sensations, will have all those characteristics which, when ascribed to sensations, worry Identity theorists. For instance, if a series of neuron-firings constitute an experience of being in pain, they

---

[3] Cf. above, p. 54.
[4] "The Concept of Mind" in Wisdom's *Other Minds*, p. 237.

must be invisible, logically nontransferable, unamenable to exact, quantitative measurement and irreducible to merely physical processes. In short, if some brain processes are sensations, then there must be something *mental* about them, in the sense outlined above.[5]

If Identity theorists admit that there are sensations in the ordinary sense of the word, then they are saddled with all the problems of dualistic theories, namely, they have the problem of trying to explain how certain properties or aspects of some brain processes are related to each other (the mental and the physical). The only alternative to their admitting sensations, as ordinarily understood, would seem to be a denial of them, but then they would be denying the very phenomena for which they are trying to account.

Scientific techniques can show, and have already largely shown, that there is an invariable correlation between certain neural processes and certain behavioral and/or mental processes. Neurophysiologists and physiological psychologists have succeeded in "mapping" many areas of the brain and in repeatedly producing certain types of mental processes by stimulating the appropriate part of the brain.[6] Extending our common sense knowledge that there is undoubtedly some sort of connection between mental and physical phenomena (drinking alcohol or smoking marijuana can affect our perception and anxiety can result in an ulcer), the empirical methods of science, including observation and experiment, have shown unmistakably that there is an invariable concomitance between neural processes of a certain sort and mental processes of a certain sort. Nobody has ever produced any scientifically acceptable evidence that any sort of mental process has occurred, or could occur, without certain neural processes occurring. Nor is it conceivable how such a claim, were it made, could ever be tested intersubjectively, since some processes or states (e.g., being in pain) have an irreducible experiential aspect of which others cannot be aware in the same way as the person undergoing them. (I need no evidence or criteria to enable me to become aware of *my* pain, but others do. Similarly, I can reveal to *you* that I have a pain, but I cannot reveal it to myself.)

It is crucial to see that even if scientists show conclusively that there is an invariable correlation between mental and neural processes, the

---

[5] Cf. above, p. 11.

[6] See Wilder Penfield, "The Nature of Speech", in *Memory, Learning and Language,* ed. by William Feindel. See also Penfield's "The Interpretive Cortex", in *Science.*

mind-brain problem will not *ipso facto* be solved. As argued before, the best that empirical techniques can do is to show *that* neural and mental processes are connected or related but not *how* they are related. Such techniques may establish which neurons are involved with specific mental processes and whether different neurons can be associated with the same mental process and vice versa. Information such as this may cast some light on the relationship between neural and mental processes but the issue of identity or non-identity is ultimately conceptual, not empirical. (We are discussing the relationship between the mental process and the neural processes which occur *simultaneously,* not that between the mental process and the neural processes which *precede* it.) If we are ever to gain a better understanding of this relationship, it will have to come from an analysis which will elucidate the status of certain mental processes and the relationship between them and the neural processes with which they are invariably correlated.

It is instructive to see how one of the world's leading neurophysiologists, Wilder Penfield, views this problem.

Let us consider the brain-mind relationship briefly....It is a boundary which, as some philosophers explain it, does not exist at all. But for the neurophysiologist there is a working boundary that does exist. Physiological methods bring him nearer and nearer to it. But he comes to an impasse, and beyond that impasse no present-day method can take him. If he should state that nerve impulses moving in certain patterns are one and the same thing as mind, he accomplishes little for his future work except to deprive himself of a useful working terminology.

Any man who adopts the dualistic terminology speaks of two elements in a living conscious human being: a body and a soul, a brain and a mind, electrical energy conducted through the *integrating pathways* of the cerebral hemispheres and *conscious thought,* a living machine and a spirit. However it is expressed, he must think either of a parallelism or a back and forth relationship.[7]

There is the suggestion here that some day an experiment might be devised whose outcome could solve once and for all the mystery of the relationship between the brain and the mind. As if we might find some phenomenon which somehow bridges the gap between the firing of neurons and an experience of pain! Then, presumably, all the gaps will have been filled, the bridges crossed and the Great Divide will no longer exist; we shall understand "how it is that nerve impulse becomes a thought and thought, in turn, electrical potential".[8]

---

[7] *Speech and Brain-Mechanisms*, p. 8.
[8] Wilder Penfield, "The Nature of Speech", p. 56.

A previous attempt to specify what such phenomena might be like is found in the writings of Descartes. His "animal spirits", as we saw, are delegated the function of carrying the mind's messages to the brain cells and thence to the rest of the body, and the body's messages to the mind. One of the difficulties with such a view is that the animal spirits are being conceived of as being spatial whereas the mind is not. Thus any commerce between the two seems impossible.

But this is the case with regard to any phenomena which are nominated as middlemen between the mind and the brain. Either they will be spatially extended and thus body or they will be unextended and thus mental.

The dilemma can be avoided only if an Interactionist view is rejected in favor of some version of mind-brain monism or a multi-aspect theory of the mind. It is the latter which is being advocated here.

So long as we, with Penfield, look upon mental phenomena as being *caused* by physical ones, we shall be puzzled about the nature of the causal mechanisms. The puzzle disappears when it is realized that there is a different relationship between brain cells and experiences than that of causation.

We have granted that there is a one-to-one correspondence between specific experiences and brain processes. But the brain processes in question are those occurring simultaneously with the experiences, not those occurring just before. When a hypnotist convinces a subject that he is in agonizing pain, the sensation and the brain processes in question occur together. But which one is "causing" the other? If they always and invariably occur simultaneously, there is no way of telling which is the cause and which the effect. Thus a non-causal interpretation is the only alternative. But this does not mean that we are obliged to adopt the Identity Theory. The fact that we can determine their simultaneous occurrence by identifying them separately indicates that the sensation cannot be "nothing but" the neural process.

Many thinkers assume that if there are mental processes, then there must be an underlying entity, thing or substance—the mind—in which or to which these processes are occurring. Such an assumption is shared by Cartesians and Identity theorists, among others.

It has long been thought that if there are mental processes which are distinguishable from physical ones, then there must be a mental entity or substance, i.e., the mind, which is distinguishable from physical ones. Such a conceptual model is, for example, at work in Hume's treatment of this subject.

In the Appendix to his *Treatise*, Hume despairs of ever finding a substance underlying his mental processes. With regard to the mind he says that *"we have no notion of it, distinct from the particular perception"*.[9] On the following page he admits that "all my hopes vanish when I come to explain the principles that unite our successive perceptions in our thought or consciousness". Earlier in the *Treatise* he describes the mind thus:

> The mind is a kind of theatre, where several perceptions successively make their appearance; pass, repass, glide away, and mingle in an infinite variety of postures and situations....The comparison of the theatre must not mislead us. They are the successive perceptions only, that constitute the mind....[10]

It is not surprising that Hume does not succeed in finding a mental entity or substance. The possibility of such a substance existing is ruled out by him because the idea of such a substance is not preceded by, and therefore based upon, any antecedent "impression":

> As every idea is derived from a precedent impression, had we any idea of the substance of our minds, we must also have an impression of it, which is very difficult, if not impossible to be conceived.[11]

For Hume, then, mental events cannot require an underlying mental substance because this turns out to be an empty notion.

J.J.C. Smart also holds that if there are mental processes which are distinguishable from brain processes, then they must be composed of some sort of non-physical substance or "stuff". If, as is the case, no such substance or stuff exists, then mental processes, e.g., sensations, "are nothing over and above brain processes".[12]

This takes us right to the heart of the Identity Theory. To say that sensations are identical with, nothing but, nothing over and above, or are really *brain processes*, seems almost indistinguishable from the claim that there are no sensations after all—or at least that there is nothing *mental* about sensations. However, the claim that none of us ever really feels pain, i.e., the claim that there really are no sensations, would not be worth serious consideration.

A more charitable and likely interpretation of the Identity Theory is that sensations do occur but are composed of and therefore *are* nothing

---

[9] *Op. cit.*, p. 559. For a lucid criticism of Hume see Terence Penelhum's article on "Personal Identity" in the *Encyclopedia of Philosophy*, ed. by Paul Edwards.

[10] *Op. cit.*, p. 229.

[11] *Ibid.*, p. 211.

[12] Smart, "Sensations", p. 163.

but neural processes, as opposed to being constituted by a mental substance or process. This is the view of Place and Smart—that sensations are not made of "mind-stuff" but of physical stuff.

Place regards the sensation-brain-process case as an example "of a special variety of the identity of things in which an identity is asserted between a state, process, or event and the micro-processes of which it is *composed*".[13] Smart argues in the following way:

> The dualist cannot really say that an experience can be *composed* of nothing. For he holds that experiences are something *over and above* material processes, that is that they are a sort of ghost stuff. (Or perhaps ripples in an underlying ghost stuff.) I say that the dualists' hypothesis is a perfectly intelligible one. But I say that experiences are not to be identified with ghost stuff but with brain stuff. This is another hypothesis, and in my view a very plausible one.[14]

Identity theorists assume, then, that unless sensations are identical with, *because* composed only of, brain processes, they must be something "over and above" brain processes and be composed of something else.

If there are mental processes over and above neural processes, then, according to the Identity theorists, there must be something in which these processes occur. And if mental processes are not material processes, then their substratum must be immaterial. And the same considerations hold, presumably, if we call mental processes 'aspects' or 'properties'. A property cannot exist on its own but must be a property *of* something. If, therefore, there are mental aspects or properties, there must be a mental entity or substance. But this means that we are back with Cartesian dualism and there is an unbridgeable gap between extended physical processes and unextended, immaterial, mental processes. Therefore, Identity theorists conclude, sensations could be nothing over and above brain processes.

It is a mistake, however, to assume that mental events, processes or aspects presuppose a mental substance. An analogy will now be developed in some detail which, it is hoped, will serve to substantiate this claim.

To assert that the weather changes when meteorological events transpire does not commit one to the view that the weather is any sort of *substance*. To deny that the weather exists would be to deny the occurrence of any meteorological events. The weather is the condition of the atmosphere at a given time and place with respect to meteorological

---

[13] Place, "Materialism", p. 102, n. 5, emphasis mine.
[14] Smart, "Sensations", p. 170, emphasis mine.

events. It is true, of course, that only spatiotemporal events comprise the weather, i.e., that the weather is "composed" of physical events in a given sequence or pattern. But it is important to note that the weather itself could not sensibly be said to be a *substance,* in the sense in which rain or snow is a substance. Nor is the weather reducible to any one of its constituents, since it is a composition, pattern or complex of these events. It is, in other words, a whole which is not reducible to any of its parts.

If meteorological events do not require an underlying substance or entity called the weather, then mental events or processes may not require an underlying substance called the mind.

Take the example of someone suffering from a toothache. Place and Smart would have it that the experience of pain which the person is undergoing is identical with, or nothing but, certain processes which are occurring in his brain. If there is something peculiarly *mental* about what is transpiring, they assume, then it must be occurring in some ghostly, mental substance called 'the mind'. Since we all know by now that such a substance is a myth, it is concluded, experiences can be nothing really but brain processes.

An analogous claim about the weather would be that since there is no meteorological (mental) substance called "the weather" (the mind), the weather is nothing really but the rain which is currently falling. And both of these claims remind us of the claim that tables which common sense deems solid are really mostly empty space since they are composed of, and are therefore reducible to, countless atoms (which are mostly empty space).

The Identity Theory refers to brain processes instead of elementary particles and it concerns consciousness or sensations instead of tables, but its message is the same in form: What common sense seems unmistakably to dictate to us, namely, that tables are solid or that experiences are in some sense *mental,* is in reality mistaken. Science reveals, the argument continues, that tables are not *really* solid and that experiences are not *really* mental. Though, admittedly, tables *appear* to be solid and experiences *appear* to be mental, we must reject the appearances in favor of the reality. The conclusion of this line of argument is that there are really fewer entities or properties in the world than common sense allows. Were we to see the world aright, were we to see things in themselves, the suggestion seems to be, it would appear very differently from what it does to the unscientific eye.

Now it seems to me that this line of argument embodies several misconceptions about the nature and purpose of scientific explanation and about the notions of appearance and reality. Only a detailed and comprehensive examination of these misapprehensions will enable us to see the Identity Theory for what it really is—an old-fashioned, reductionist piece of metaphysics giving the appearance of a novel scientific theory.

When Identity theorists identify consciousness and sensations with brain processes, they think they are not only *accounting* for these phenomena but they are drawing our attention to the real nature of consciousness and sensations. Consciousness and sensations, they are saying, are as accessible to us as any other phenomena in the world. Science is adequate to account for any phenomena, even those which have hitherto been regarded as somehow private and personal (though the person undergoing them can, and often does, disclose, reveal or exhibit their presence to others). J.J.C. Smart voices this confidence:

That everything should be explicable in terms of physics (together of course with descriptions of the ways in which the parts are put together—roughly, biology is to physics as radio-engineering is to electromagnetism) except the occurrence of sensations seems to me to be frankly unbelievable.[15]

Sensations are simply physical processes, Smart thinks, because there "does seem to be, so far as science is concerned, nothing in the world but increasingly complex arrangements of physical constituents".[16]

Thus, processes which are in principle not publicly observable would not lie within the purview of science and hence must be rejected, Identity theorists argue. Materialists are committed not only to the denial of privately detectable substances but also to privately detectable processes, states or events.

It can be argued that pains are not necessarily private states or processes because they can be, and characteristically are, exhibited, revealed or disclosed by the organism in pain. And, I submit, this argument is valid. Pains are at best contingently private. However, the fact remains that I could not and need not appeal to evidence or criteria to determine whether I am in pain, whereas others can and must make use of such data. And this would still be the case if others could somehow observe all the physical processes occurring in me before and during my pain.[17]

---

[15] Smart, "Sensations", p. 161.
[16] *Ibid.*
[17] Cf. above, pp. 54-55.

We are forced to conclude that there are some features of the world which are not intersubjectively, i.e., publicly observable. In the case of at least some of these phenomena, e.g., consciousness and sensations, it is probably misleading to say that they are observable in *any* sense, other than the sense that one *undergoes* them. In this sense, to say that one "observes" one's sensations but others cannot, means neither more nor less than to say that one has or undergoes one's sensations, but others cannot.

This does not mean that sensations are in every way ultimately private. It does not mean that others cannot be apprised of their occurrence. But the fact that others need to be told about one's pains or to see signs of them means that at times we can refrain from telling them and can suppress the signs. This is one of the facts about pains which leads us to say that they are, or can be, in some sense private. But there are natural ways in which pain manifests itself or is expressed and the more severe the pain is, the more effort it takes to suppress these natural signs, e.g., moaning or writhing.[18]

Not all Identity theorists deny the existence of mental processes which are private in some sense. For instance, Herbert Feigl writes:

....it makes perfectly good sense to speak of the subjectivity or privacy of immediate experience. *Numerically* different but *qualitatively* identical (indistinguishable) experiences may be had by two or more persons, the experiential events being "private" to each of the distinct persons.[19]

He adds that though a doctrine involving the notion of "absolute" privacy may have cognitive meaning, he is "inclined to regard it as scientifically meaningless".[20] That is, for Feigl, science could make no use of contents which are absolutely private. Presumably he means that if something is "absolutely" private, it could not even be communicated to others whereas if it were "relatively" private, so to speak, it could be disclosed or communicated and perhaps be utilized in scientific experimentation.

Feigl says that it is absurd to deny the existence of raw feels (sensations) as materialists do[21] and he elsewhere says that radical materialists "typically *repress* or *evade* the mind-body problem. They do not offer a genuine solution".[22]

---

[18] Cf. below, pp. 135-36.

[19] "The 'Mental' and the 'Physical' ", p. 398.

[20] *Ibid.*, p. 402.

[21] *Ibid.*, p. 429.

[22] "Mind-Body, *Not* a Pseudoproblem", in *Dimensions of Mind,* ed. by S. Hook, p. 29.

But Feigl does claim that sensations are identical with neural processes.

One reason why Feigl and Smart seem to be attracted to the Identity Theory is that they deem certain mental processes to be outside the causal chain of events. The physical universe is a closed system, they argue, and the physical laws of the natural sciences can theoretically account for everything that occurs in nature. Feigl, in criticizing Epiphenomenalism, says the following:

Epiphenomenalism....accepts two fundamentally different sorts of laws—the usual causal laws and laws of psychophysiological correspondence. The physical (causal) laws connect the events in the physical world in the manner of a complex network, while the correspondence laws involve relations of physical events with purely mental "danglers". These correspondence laws are peculiar in that they may be said to postulate "effects" (mental states as dependent variables) which by themselves do not function, or at least are not needed, as "causes" (independent variables) for any observable behavior.[23]

In other words, private, causeless, mental processes would be "messy". They would be unaccountable in terms of physics or chemistry. They would be inaccessible to science if they could only be "observed" privately. And they would be inexplicable in terms of evolutionary theory since they would be utterly useless (causeless) in terms of survival. Science, in other words, cannot account for such processes, so they had better be identified with the neural processes which occur simultaneously with them or else interpreted in terms of dispositions.

Though Feigl regards Smart as an ally,[24] it is doubtful whether Smart would reciprocate unqualifiedly. This is because Feigl says that the common referents of neurophysiological and phenomenal terms are "the immediately *experienced* qualities, or their configurations in the various phenomenal fields"[25] whereas Smart holds that the common referent is the neural process.

I have emphasized the word 'experienced' here for a good reason. Because one does not experience, i.e., is not aware of, one's own neural processes (e.g., the firing of certain neurons) when one experiences pain, the Identity Theory strikes one as being *prima facie* absurd. Surely, if sensations were nothing but neural processes, one would be aware of one's neural processes when one is aware of, i.e., has, one's sensations. Since this is not the case, it is difficult to discover exactly in what *sense*

[23] *Ibid.*
[24] *Ibid.*, p. 35.
[25] *Ibid.*, p. 31, emphasis mine.

sensations could be identical with neural processes. A person need not observe his own behavior or learn about his own neural processes in order to determine whether or not he is in pain. But others need to observe his behavior or his physiological processes if they are to find out that he is in pain. Both of these facts indicate that one's being in pain could not consist merely of behavior and/or of neural processes. If pain were nothing but behavior and/or neural processes, then one could *investigate* whether or not one were in pain, which makes no sense, and others could know that a person were in pain before *he* did, which is absurd. These epistemological facts about pain, then, reveal a *prima facie* implausibility in the Identity Theory which for many vitiates it as a live option.

In a recent article, Isabel C. Hungerland suggests that a person in pain 'does' observe his own pain-behavior and that this observation is not irrevelant to his having a right, or being able, to ascribe pain to himself. This is how she depicts "an uncomplicated paradigm example of *being in pain*":

The whole experience is something like this. One has an extraordinarily distressing feeling, located roughly in a relatively large area of one's body, and one both feels and observes the natural manifestations of distress, of unpleasantness that cannot be avoided or run away from because it is located in our own body. One feels the scream "rising" in one's throat and hears it as one hears another's. One feels and observes the writhing limbs, the flinching torso, and so on.[26]

Professor Hungerland adds the following a little later:

For self-ascribings there is, as I made out in the paradigm case, as a basis for support, both "having" the feeling and feeling the manifestations, as well as observing them. In other-ascribing, there is and could be only observing of manifestations.[27]

We find here the suggestion that it is necessary, or at least relevant, for one in pain to be aware of the signs of his pain and to observe his own behavior in support of his ascribing pain to himself. Such a suggestion is often made by those who are working on the problem of Other Minds. Ryle, as we noted before, argues that self-knowledge is to all intents and purposes on a par with knowledge of others. It is my contention, however, that observing the signs of one's own pain or observing one's own behavior is utterly irrelevant either to the "discovery" that one is in pain or as a support for one's claim to being in

---

[26] "My Pains and Yours", in *Epistemology,* ed. by Avrum Stroll, pp. 120-21.
[27] *Ibid.,* p. 123.

pain. Thus knowledge of one's own sensations is not on a par with, is not obtained in the same way as that of, knowledge of the sensations of others.

A paradigm case of being in pain is found when we consider what occurs when one is having a cavity filled at the dentist's. The patient reclines with his eyes tightly closed, perhaps on the premise that what he doesn't see won't hurt him. If he has not had any novocaine, or its equivalent, he begins to feel pain whenever the drill comes close enough to a nerve. And, if the drill actually touches a nerve, the pain is instantaneous, unmistakable and excruciating. True, one may be aware during this time that one has tightened one's grip on the arms of the chair, that one's whole body has stiffened, that one is moaning loud enough for the dentist to hear and that one's eyes are beginning to fill with tears. But none of these data is necessary or relevant either for the sufferer to determine that he is in pain or as a support for his covert or overt self-ascriptions of pain. It is perfectly conceivable that none of these phenomena will be noticed, either because none of these things occur or because the patient is aware of nothing but the pain which seems to be filling his whole head. He will be aware of his pain and can legitimately ascribe pain to himself, without noting anything else but the pain.

The dentist and his assistant, on the other hand, will know that their patient is in pain, and legitimately be able to ascribe pain to him, only if he expresses his pain verbally and/or bodily. Both Ryle and Hungerland realize this, but they seem to think that my overhearing my own moans and my observing my own bodily reactions to my pain somehow makes possible or enhances my knowledge of the existence of my pain.

All of this is relevant to the claim that sensations are nothing but neural processes. If this claim were true, then others could, in principle, determine whether or not the patient in the above example were in pain at precisely the time at which he does. But if sensations were nothing but brain processes, then it is difficult to conceive how the sufferer himself *discovers* that he is in pain. It is not, after all, common for people to know what is transpiring inside their brains, yet they seem not to find it difficult to determine whether or not they are in pain. Were sensations nothing but brain processes, it would make sense to say that for a person to discover whether or not he is in pain, he must first observe, as others must, his brain processes and then *infer* that he is in pain. But this is patently absurd—he would not be motivated to observe his brain processes (in an autocerebroscope, for example) unless he *were* in pain, in which case he would *not need* to observe them.

One need not observe or overhear oneself in order to determine that one is in pain, but others must. And nothing would be changed if we tentatively adopted the Identity Theory. It would still be necessary for others to observe, let us say, a person's neural processes, before they could know that he is in pain. In fact, if sensations were nothing but brain processes, only those who know of the occurrence of the latter would know that someone is in pain.[28] But this is simply not the case.[29] If we must first learn of the occurrence of certain neural processes *in order to ascertain* whether someone is in pain, this in itself shows that the neural processes and the pain cannot be simply identical. They must be different processes or different features of some composite process or perhaps the same in one sense but different in another sense.[30] They cannot simply be identified with each other.

It may be said that to deny an identity here forces us back into Cartesian dualism. To say that pains are *different* from neural processes suggests that they exist as ghostly processes in a gaseous medium too subtle to be detected by any of the five senses or to be microphotographed. Or, it may be interpreted as the claim that pains are different *sorts* of things from physical processes and that therefore they must occur "in the mind" and not in the body. But to deny that pains are different from neural processes may be interpreted as the claim that they are not *mental* processes. Thus we seem to be faced with the choice of postulating ghostly, undetectable entities or denying the obvious fact that people sometimes feel sensations and that there are certain epistemological facts about these phenomena which lead us to call them 'mental' and to insist on an asymmetry here between self-knowledge and knowledge of others.

It is possible to deny an identity between sensations and brain processes without postulating a gaseous mental medium. And an alternative to the Identity Theory need not postulate mental entities in addition to physical entities, stipulating that the mental ones are ghostly or ethereal. A Multi-Aspect Theory of the mind will be presented and defended in the course of this discussion. This theory represents an attempt to avoid both Cartesian dualism and materialistic reductionism.

[28] Cf. immediately above and n. 31 p. 156 below.
[29] Cf. above, pp. 54 ff.
[30] Cf. below, p. 76.

CHAPTER FOUR

REDUCTION AND REALITY—
SOME MISCONCEPTIONS ABOUT SCIENCE

It was argued earlier that the experiential feature of pain—the feeling of pain with which we are all familiar—could be said to be in the same logical category as its mental features. What this means is that Ryle is mistaken when he identifies the mental with dispositions. Sensations are not dispositions—though they may involve the having of certain dispositions—but are occurrent processes, or occurrent features of processes. In this sense they are on a par with, are in the same logical category as, neural processes. Both are ongoing phenomena as opposed to being dispositions.

To say that someone is in excruciating pain, says a different sort of thing about him from saying that he is brilliant. The ascription of pain to someone means that he is currently in a certain state which, we now know, incorporates both experiential and neural features. States of excruciating pain normally involve numerous bodily and behavioral features also, aspects of the state which can serve as signs to others that the individual in question is in pain.

Being in pain, then, unlike being intelligent or vain, primarily involves being in a certain state as opposed to being merely disposed to feel or able to act in certain ways. But having pains and being intelligent have something in common which neither shares with neural processes, namely, the concepts of undergoing experiences and of being intelligent are both features of our concept of mind. The concept of mind involves both dispositions and occurrences. Or, in other words, we unhesitatingly ascribe a mind to something only if it is at least sometimes conscious or aware of something and if it displays some degree of intelligence.

Thus it is misleading to say, with Descartes, that the mind and the body are disparate substances or, with Ryle, that the mental and the physical are in different logical categories, the former being largely a set of dispositions. Being in pain is a state with neural and experiential features and usually others as well, e.g., characteristic behavioral manifestations.[1] Both Cartesian Dualism (which at times identifies a person with his consciousness) and Rylian Dispositionalism fail to account for, or even to accept, man's duality. But it is misleading even to say "duality" here, because this is reminiscent of the view of man as an anomalous amalgam of an unextended Cartesian soul (mind) and an extended body. The fact is that there are as many dimensions, features or activities of man as there are different sorts of ascriptions of him. A man can have a wound, have neural processes, have a pain, have an idea, have superior intelligence and have a game of golf or chess. It is implausible to say that every one of these events, states, processes, dispositions or activities can be neatly classified as simply mental or physical.

We have noted that the Identity Theory has sometimes been regarded as primarily scientific. Smart suggests that science may soon reveal the "true nature" of sensations,[2] namely, that they are really composed of, identical with, and so nothing but, brain processes.[3] Place also suggests that sensations are to be identified with the brain processes with which they are invariably correlated, or of which they are composed,[4] though he distinguishes between the factual discovery of these correlations and the logical inference to an identity.

We shall now examine more conceptual issues raised by the Identity Theory. These are by far the most important issues since the Identity Theory is really more a philosophical claim about "reality" than it is an empirical thesis which attempts merely to *account* for sensations.

For the purpose of argument, I have adopted the position that in the future, neurophysiological experiments will have demonstrated an invariable correlation between sensations and brain processes. Unless there is such a correlation, the Identity Theory is utterly untenable. We shall assume, further, that it makes sense to say, and has even been established that, sensations are *composed* of neural processes, as Place and Smart claim. Even granting this, however, in no way establishes the Identity Theory, i.e., the thesis that sensations are nothing but, are simply identical with, or are really, neural processes.

It is as if one were to say that tables and chairs are composed of elementary particles, e.g., protons, neutrons and electrons, and that there-

---

[1] For a discussion on why pain is classified as a 'mental' state, see pp. 142. ff.
[2] Cf. above p. 46.
[3] Cf. above, pp. 41 and 63.
[4] Cf. above, p. 63.

fore tables and chairs are really nothing but, or are simply identical with, these elementary particles. Ernest Nagel argues as follows:

Scientific method is largely concerned with the analysis of objects into their constituent elements. Thus, the physicist, the chemist, the geologist, and the biologist each seeks to find the constituents of the objects he studies; psychology, the social sciences and philosophy try to do the same. It is understandable, therefore, how the misconception arises that science identifies objects with their *elements*. Science, however, does *not* do so, but analyzes its objects into *elements that are related* to each other in certain ways, so that if the same elements were related in different ways they would constitute *other* objects.[5]

Thus it is a misconception of science to interpret it as in some sense *reducing* objects to their ultimate constituents. It makes no more sense to say that tables and chairs are really nothing but the chunks of wood of which they are composed, than to say that they are nothing really but the elementary particles of which they are composed. If these claims made any sense, then the question would arise whether tables are really nothing but chunks of wood *or* molecules *or* atoms *or*.... The absurdity of this question, and the infinite regress to which it gives rise, shows the absurdity of the view that a scientific account of a phenomenon somehow *reduces* that phenomenon to the entities cited in the account.

Thus, assuming that it makes sense and is true to say that sensations are composed of neural processes, it does not follow that they are identical with, or reducible to, or are *really* just these neural processes. Sensations may *be* neural processes and yet not be reducible to them. If they are not ontologically reducible to them then they are not reducible to them in any other sense, e.g., epistemological or linguistic. I argue that *because* they are not reducible ontologically, they are not reducible in the other ways. Sensations might *be* neural processes arranged or related in certain ways and occurring in a living organism, but they are no more reducible to the neural processes than are words reducible to the letters which constitute them.[6] What we have here is a case of S being the same as B yet different from B. It is this relationship which I try to elucidate throughout this discussion.

It might be argued that if sensations are composed of neural processes, then materialism is true. This would be so because what were formerly regarded as mental processes are really just physical processes. However, it will be shown that there is nothing implausible in the view that different

---

[5] Nagel, *Introduction to Logic*, pp. 382-83.
[6] Cf. below, pp. 74 ff.

sorts of predicates (e.g. 'mental') can be made of certain phenomena which cannot be made of their constituents considered individually or as a mere sum. These phenomena may possess certain properties—certain facts may obtain concerning them—as a result of the way in which their constituents are arranged and/or the way in which the latter may interact with each other. Different properties, even different *sorts* of properties, can and do emerge from particular combinations of the constituents of the phenomena which will be discussed. Sensations may be and in fact are, physical phenomena, but they have properties—there are facts about them which we have seen—which legitimate our calling them 'mental' too.

Saying that sensations are identical with, or are nothing but, neural processes, is not unlike saying that statements are nothing but words or that words are nothing but letters (or that letters are nothing but marks on paper, etc.). The expression "are nothing but" is used in such claims as a euphemism for "do not really exist" or "are not real"—in the sense in which their constituents are 'real' or 'really exist'. This, surely, is exactly what Identity Theorists are claiming about sensations. This is what their identity of composition means. Most Identity theorists (Feyerabend is an exception) [7] are reluctant to admit that they are denying the existence of sensations as these are ordinarily conceived, but this is a plausible interpretation of their theory.

What Identity theorists are saying, in short, is not just that the occurrence of sensations can be fully *explained* by reference solely to neural processes, but that sensations are *really* neural processes. And this claim is no different in kind from the claim that water is *really* oxygen and hydrogen atoms or the claim that tables are *really* a certain type of molecule or the claim that words are *really* letters.

Identity theorists, at least at times, claim to be advancing a scientific theory which accounts for, and illuminates the nature of, sensations, just as the atomic theory has enabled us to account for, and illuminate the nature of, material objects. However, as Nagel indicates, scientists do not make the claim that, for example, water is really (i.e. nothing but) hydrogen and oxygen atoms but the claim that water is composed of these atoms arranged in a certain way. It is as unilluminating to say that water is "really" just two atoms of hydrogen and one of oxygen (and

---

[7] "...if you want to find out whether there *are* pains...in the sense indicated by the common usage of these words, then you must become (among other things) a materialist", Paul Feyerabend, "Materialism and the Mind-Body Problem", *Review of Metaphysics*, p. 53.

not what we thought it was before the atomic theory was devised) as it is to say that sensations are "really" neural processes (and not what we thought they were before the neurophysiological facts were known). And these claims are as illuminating as the one about words being "really" letters.

In a sense, words *are* nothing but letters. That is, there is a sense in which they are identical with, nothing but, or composed of letters. But there is a more important sense in which they are *not* nothing but letters. Let us take as an example the word 'form'. It is composed of four letters. But it would be misleading to say that it is *identical* with, or nothing but, these four letters. For the word 'from' is also composed of the same four letters. If 'from' were identical with the four letters and 'form' were identical with the four letters, then 'from' would be identical with 'form', which it is not. Yet two things which are identical with a third thing must be identical with each other.

Words are formed from, or 'emerge' (if you will) from, particular relations or combinations of letters—excepting words consisting of only one letter, e.g., 'a' and 'I'. Words can have different properties from those of letters. They are often defined as "the smallest unit of meaning". Letters are not said to have meaning, except when they constitute a word like 'a' or 'I'. Words may be apt, ill-chosen, or "dirty" but letters could be none of these things. Words are a different *sort* of phenomenon from letters. They serve a different purpose and they have different properties. Words are not just a sum of letters. The crucial thing is the way in which the letters are arranged and, of course, the way in which the word is used in the statement of which it is a part.

It is also not accurate to say that statements (or sentences) are identical with the words which constitute them. For example, the sentence "Victoria loves Nigel" is not the same as the sentence "Nigel loves Victoria", yet each is constituted by the same three words. It would be highly misleading to say that either sentence was identical with, or nothing but, the three words because, as is obvious, the words can constitute quite different sentences. As with the letters, so with the words—the arrangement is the crucial thing. Also, how the sentence is spoken in conversation can determine its meaning. This is another indication that statements are not simply identical with the words which constitute them. The same words can be used to make different statements, depending on the circumstances, the tone of voice and the emphasis placed on one or another of the words.

It would be very interesting to find out whether the same neural processes are involved in the experiment where the word 'pain' appeared and the one in which it was omitted.[8] If the same neural processes can be involved in different experiences of pain, this would indicate that the experiences could not simply be identified with the neural processes.

Statements have properties which neither letters nor words do. They may be true or false, for example, but neither individual letters nor individual words could be literally true or false (except in cases where a single word can constitute a statement, e.g., "Yes" in answer to a question). Statements are different sorts of things from words, which are different sorts of things from letters. All three are, of course, equally real even though, *in a sense,* statements are "nothing but" words and words are "nothing but" letters. As we have seen, however, it is the sense in which statements are not just words, and words are not just letters, that is the interesting and important sense.

Anyone who answered the question "What are statements?" by saying "Statements are identical with words" or the question, "What are words?" by saying "Words are identical with letters" would be saying something that is very trivially true, but is utterly inadequate, highly misleading and totally irrelevant. The same holds of those who think they are adequately characterizing sensations by saying that "sensations are brain processes". If pains are neural processes, then they are neural processes only in the sense in which statements are words. Pains are a different sort of thing from neural processes and they have different properties than the latter have. If there is nothing implausible in the claim that statements are more than just words or the claim that words are more than just letters, there should be no implausibility in the claim that sensations are more than just neural processes.

It may be argued that there are some cases where the whole is nothing beyond its parts and that perhaps conscious states and neural processes are in this category. For example, a wall composed of bricks and a Cabinet composed of Cabinet members may be nothing beyond the parts which constitute them.

I do not object to the statement that a wall *is* the bricks which constitute it. But what is meant by the claim that a wall is "nothing beyond" or "nothing but" the bricks? The same bricks, after all, might have been used to make a bookcase. The same men who form the Cabinet might have constituted a football team. It is not the identity which is suspect in these cases but the 'strict' identity which is implied. I do not wish to

---

[8] Cf. above, p. 35.

suggest that the wall or Cabinet is *more real* or something *over and above* its constituents. I submit, however, that no whole is *reducible* to its parts though of course it will be composed only of its parts. To say that a wall is "nothing but" or "nothing beyond" its bricks is to imply that the whole is unreal or is at least somehow less real than its parts. This is the implication of the Identity theorists which must be resisted. There would be no *point* to the Identity Theory if it amounted to the claim that sensations are constituted by neural processes.[9] If Identity theorists admitted that neural processes sometimes constitute sensations, just as bricks or Cabinet members sometimes constitute a wall or a Cabinet, then their theory would lose its *raison d'être*.

Perhaps another analogy will help. A chocolate cake may, in a sense, be identical with the items listed in a recipe. But how different the cake may turn out when made by different cooks, even though exactly the same ingredients are used! Much depends upon the order in which the ingredients are added, how they are mixed, the type of utensils used and the length of time it is left in the oven, etc. What I am suggesting is that the whole may be much more than, or something different from, the sum of its individual parts. I am also suggesting that very different properties may be produced by varying the arrangement and/or the interaction of the constituents. A cake can have different properties from its individual constituents. Its taste, for example, may be utterly unlike any of its ingredients. So may its shape and smell.

It may be objected that these analogies can throw no light upon the case of sensations and neural processes because the human brain is infinitely more complicated than letters of the alphabet or the ingredients of a cake. But this is precisely what I wish now to emphasize. The more unlike these phenomena (letters, words, etc.) the human brain is, the easier it is to conceive that some of its processes might well constitute phenomena with properties unlike any or all of them.

It must be remembered, too, that the human brain is by far the most complicated phenomenon yet encountered. It is not surprising that its functioning should result in properties, processes or effects which are to be found nowhere else in nature. If letters arranged in the appropriate way can result in a word, if words arranged in a certain way can result in a statement, and if the ingredients listed in a recipe can be so manipulated that they result in a culinary masterpiece, surely it is not implausible that

---

[9] To give the Identity Theory its due, it does rightly criticize the notion that sensations are ghostly processes composed of some ethereal 'stuff'.

the arrangement and interaction of millions of neurons could constitute a sensation of pain with certain properties unlike those of its constituents.

Identity theorists think that once all the causal conditions or constituents of sensations have been found and we can account for sensations, then they must in reality *be* just these conditions or constituents. An Identity theorist with regard to statements or cakes would claim that they are *nothing but* the ingredients which constitute them. But once we see how a whole can *be* its parts without being reducible to them, the Identity Theory loses its *point,* viz., that sensations are not what we have always thought them to be but are *simply* neural processes.

Feyerabend realizes that sensations appear to be very different sorts of things from neural processes but claims that we should not be deceived by appearances. In his paper "Materialism and the Mind-Body Problem", he says the following:

...what appears to be different does not need to *be* different. Is not the seen table very different from the felt table? Is not the heard sound very different from its mechanical manifestations....? [10]

In common with many other Identity theorists, e.g., Smart, Feyerabend identifies heard sound with its mechanical manifestations. Smart interprets perception of colors as nothing but discriminative behavioral responses. Armstrong interprets *all* perception as the acquiring of actual or potential belief (information).

Identity theorists seem to give with one hand what they take away with the other. For example, Smart says "I do not wish to deny, of course, that a man who sees a Cambridge blue oar has a different sort of *visual experience* from that had by a man who sees an Oxford blue oar".[11] However, he adds immediately, "visual experiences are in fact brain processes".[12]

Thus Identity theorists seem to acknowledge that we have experiences, e.g., hear sounds, see colors and feel pains but they wish to identify these experiences with brain processes. Not only Feyerabend, but also Smart, seems to think that we have all along been unaware of the *real* nature of sensations—in *reality* there are no sensations apart from the firing of certain neurons. Smart makes the following claim in *Philosophy and Scientific Realism*:

[10] *Op. cit.,* p. 54.
[11] "Colours", in *Philosophy,* pp. 138-39, emphasis mine.
[12] *Ibid.,* p. 139.

....it may be the *true nature* of our inner experiences, as *revealed by science,* to be brain processes....[13]

Identity theorists are not claiming that our experiences are *also* neural processes; they are claiming that they *are* neural processes. But if this identity is to be revealing in some way, if it is to inform us of anything, then we must somehow have all along been mistaken in thinking that we knew what, e.g., pains were. What, then, could it mean to say that pains may be different sorts of things from what we have all along thought them to be? When Smart claims that science may reveal the "true nature" of pains to be brain processes (instead of what?) he is conflating either the causal conditions of pain, or the physical constituents of pain, with the sensation of pain itself. What he is saying, in effect, is that once we know what causes pain, or what are its neural components, we will realize that pains are nothing but these causes (causes of what?) or constituents (constituents of what?).

Perhaps it will be objected that the mere fact that we refer to mental items, e.g., pains, and describe them in certain ways, e.g., 'excruciating', does not in itself render the Identity Theory untenable. After all, the mere fact that people *talk* of demons, witches, miracles or flying saucers does not preclude our explaining away these phenomena by philosophical analysis and/or by scientific theory. Perhaps there really are no sensations, despite what people do or do not *say*. Pains will go the way of demons, witches, miracles and flying saucers if we discover that we have been mistaken all along.

It seems to me that this objection presupposes a conception of mind and mental events which is insupportable. The presupposition of this objection is that a pain must be either of two things and nothing else, i.e., a physical event *or* an immaterial event. Either the mind is the brain and/or behavior or it is an immaterial substance which is undetectable by scientific or intersubjective means.

Because Identity theorists recognize only these two possibilities, i.e., Materialism or Cartesian Immaterialism, they feel that we must once and for all face the fact that sensations and all other supposedly mental states are simply *physical* states or we are forced to admit the existence of ghostly items, e.g., pains, whose ontological status is in fact no different from that of demons, witches, miracles or flying saucers. Mental states are either physical states, as are all other states in the universe, or they

---

[13] *Op. cit.,* p. 93, emphasis mine.

must be a queer sort of insubstantial event which forever eludes our scientific instruments of detection. Take your choice—*physical* events or *ghostly* events.

I have argued throughout this discussion that this conception of mind and mental events is subject to the same criticisms as the Cartesian conception. To argue or to presuppose that the mind or mental events must be either physical *or* mental (ghostly) is to miss the point of much of this whole discussion.

We have argued that there are at least two broad categories of the mental—behavior and experience (including consciousness).[14] We have, in the main, largely ignored behavior and have concentrated our attention upon pain as an example of a conscious state. We have argued that there are various criteria of mentality and that sensations may be composed of neural firings without being reducible to those processes. Thus both behavior and states of consciousness can be both physical *and* mental.[15] The dichotomy "mental *or* physical" is thus untenable.

To suggest that pains may go the way of demons and witches thus presupposes that pains must be *nothing but* neural firings or else they can only be ghostly events occurring in a ghostly medium (the mind). I have rejected the first of these alternatives and have argued that we need not have recourse to the second alternative. We need not be either Materialists *or* Cartesians.

Also, I have argued that it is not just "mere talk" when we refer to pains. We could stop calling people 'witches' and we could stop calling events 'miracles', but could we stop calling what we feel 'pains'? I have argued that even if we substitute 'Materialese' for 'Mentalese', we would still call pains 'mental'. Whether pains are neural firings, immaterial events, or something else again, we shall continue to call them 'mental' for the reasons I have outlined earlier.

So long as it *hurts* when one is struck or burnt, etc., so long will we require the notion of pain as a mental event. One could never show this hurt or pain to another but this is not because it is an immaterial, invisible, ghostly event.[16] We have conceded that pains may be composed of nothing but neural firings yet insist that they are still mental, in all of the senses outlined above.

---

[14] Cf. above, pp. 10 ff. and 14 ff.

[15] In *The Concept of Mind*, Ryle shows that behavior can be *both* physical and mental.

[16] Cf. below, p. 139 on the independence of bodies.

We do not infer from the "mere fact" that we talk "Mentalese" that there must be mental events. That would be like inferring from our talk about demons and witches that there must be demons and witches. We have tried to show in detail what is mental about pains and other such states. It may be that for most people the word 'mental' means what it does to Materialists and Cartesians alike, i.e., unextended, invisible, undetectable by any of the senses, ghostly, ethereal, separate from the body, necessarily conscious, etc. Presumably a poll would have to be conducted in order to determine if this is what most people mean by the word 'mental'. Instead, I have tried to show what is involved when we call something 'mental'; this resulted in my discussions on the various criteria of mentality.

We might be able to cease talking of 'pains' and talk in terms of neural firings alone. I am not inferring from our current use of 'pain' to the existence of pains. (Is there really any need to *infer* the existence of pains?). It should be obvious, however, that so long as we continue to have experiences or sensations—whatever these are composed of— and so long as the pervasive facts about privacy and privileged access, etc., continue to *be* facts, talk in terms of neural firings alone will be 'infected' with 'Mentalese' in one way or another. For we will continue to think and feel; and so long as we do think and feel, we shall want to show and tell others about it. We agree that from the fact that people use certain words we cannot infer that these words have referents. For example, people use the word 'God' but this does not entail that there is a God. And the use of words such as 'but', 'and', 'is' or words such as 'freedom' and 'democracy' does not entail that there are referents corresponding to these words. The existence of things is determined not by using words but by such procedures as experience, observation, testing, etc.

There are some who would require intersubjective observability as a criterion of the existence of all events, states, processes or entities. This requirement precludes the existence of sensations and other states of consciousness, as these are normally understood, because they are not intersubjectively observable. But I have argued that pains need not be either intersubjectively observable *or* ghostly entities. The neural firings which constitute pain occur inside living organisms and the pain felt by the individual is the integrated or unified complex of these neural processes as they occur in him. To identify the pain with the neural firings is to identify the whole with its parts. But the whole is a result of the particular neural firings which are occurring in a specific sequence and pattern, just as a whole word is a result of particular letters occurring in a specific sequence. No matter what constitutes sensations or words, it is not enlightening, though it may be trivially true, to say that they are their constituents. This is why I reject the Identity Theory as an attempt to correctly and adequately characterize sensations or other states of consciousness.

Thinking that sensations are adequately characterized as neural firings is like thinking that words are adequately characterized as letters or that statements are adequately characterized as words.[17] The following dialogue was intended by its author as a joke but there seem to be some thinkers who would nod their heads approvingly:

Pol.　What do you read my lord?

Ham.　Words, words, words.[18]

Ought Polonius to be satisfied with Hamlet's answer? Is his answer false? Is Hamlet not reading words?

We see that Hamlet's answer is not so much false as it is irrelevant or at least inadequate. And so is the Identity Theorist's claim that consciousness or a sensation is a brain process. The Identity Theory does not provide a plausible or adequate characterization of sensations. I am not denying that it may be important to know what constitutes sensations but I am denying that knowing an item's constituents *ipso facto* amounts to knowing its 'true nature' or what it 'really' is.

What seems to attract thinkers to the Identity Theory is its negative import, i.e., its denial that consciousness or sensations are ghostly, incorporeal processes. It seems much more plausible that they are brain processes rather than elusive, insubstantial ones. And the Identity theorists are right; there are no ghostly processes in the 'machine'. And they are not mistaken when they say that sensations are neural processes.[19] But they are mistaken when they say that sensations are *nothing but* neural processes because this suggests that they are no different from any physical or physico-chemical-electrical process occurring anywhere in nature. And the fact is that pains are not just electro-chemical-physical processes. No matter what their constituents are or what makes them up or what underlies them they are and always will be experiences of

---

[17] Place seems to think that the expression 'brain process' might provide an "adequate characterization" of consciousness; see, e.g., "Is Consciousness a Brain Process?", p. 103.

[18] Polonius and Hamlet; from William Shakespeare's *Hamlet,* II, ii.

[19] Nor was Hamlet in error when he told Polonius that he was reading words.

living organisms. And to suggest that pains may be in the same category as demons, witches, miracles or flying saucers, i.e., in the category of the non-existent, is to suggest that living organisms may not have pains. This is, of course, a logical possibility, but the fact is that we do.

An Identity theorist, it may be argued, could agree with everything I have said about the need for a mental terminology and yet still hold that sensations and other mental states are nothing but neural processes. There may be many different ways of characterizing one and the same item, e.g., a table. Why not pains too? We may even *need* a way of characterizing pains as 'mental', but they could still be nothing but brain processes. That is, the existence and/or the need for a mental terminology does not in itself mean that pains are something over and above brain processes.

The question is, however, whether pains are correctly characterized *solely* in terms of brain processes. I submit that they are not—no more than words are correctly characterized solely as letters. To characterize anything solely in terms of its constituents is to *mis*characterize it. To characterize a person as being nothing but a very complicated physico-chemical mechanism or nothing but a physical object—though we may *be* one or the other of these things—is to *mis*characterize a person. Of course persons are bodies, words are letters and sensations are brain processes; but it is wrong to suggest that the latter expressions are correct or adequate characterizations of the former.

Identity theorists and their critics might agree that sensations are brain processes. But when the Identity theorists infer from this fact (if it is a fact) that they are 'nothing but' brain processes, they must be reminded that the brain processes in question do constitute a sensation. Those who advocate the 'disappearance' form of the Identity Theory [20] respond to this reminder by denying an identity between sensations and brain processes. They see no need for a mentalistic terminology and are searching for a strictly materialistic terminology which will make no reference to pains or, presumably, to any other mental states or processes. Their aim seems to be to develop a language which refers only to intersubjectively observable and testable events, i.e., a scientific

---

[20] The suggestion is that when science has revealed what pains really are, we will cease to believe in pains just as we have ceased believing in demons; see, e.g., Richard Rorty, "Mind-Body Identity, Privacy, and Categories" in *Philosophy of Mind,* ed. by Stuart Hampshire, esp. p. 37. See also Paul Feyerabend, "Comment: Mental Events and the Brain", in the *Journal of Philosophy.*

language which could, in theory, replace the natural languages of man. This new language, in which misleading subjective utterances such as "I'm in pain" will be replaced by accurate and objective utterances such as "My C-fibers are firing", will be the language of "total science".[21]

I have argued that adoption of new terms to refer to our pains and other sensations would not solve the mind-brain problem.[22] Of course "My C-fibers are firing" might come to convey what "I'm in pain" conveys now. But this will not alter the fact that pains cannot be inter-subjectively observed, even though their constituents may be. Perhaps we could see a person's C-fibers firing but what this feels like to him could never be photographed or examined under a microscope. Whether his feeling is composed of ghostly or neural processes, or of something else or of nothing, it could not be viewed in a laboratory or anywhere else.

It seems to be one of the presuppositions, if not an explicit claim, of Identity theorists that there is "nothing in the world but increasingly complex arrangements of physical constituents".[23] Were this the case, critics of the Identity Theory could argue that there is *no need* to criticize it because it does not exist. How, after all, could there be any theories if there were nothing in the world but the elementary particles in various arrangements? To defend a strictly materialistic *theory* is to deny the existence of what one defends. If materialism is true, then it is *nothing but* marks on paper or sounds (sound waves?) uttered by human beings. Since fruitfulness and inutility apply to theories in a different sense than to aggregates of matter, then if materialism is fruitful or true, it is neither fruitful (true) nor false. Since this is a contradiction, materialism cannot be true.

Theories may be *about* aggregates of matter but they themselves are a different sort of thing. Materialists claim that all the apparently different sorts of things in the world are really only one sort, i.e., material. However, even among the elementary particles there are different sorts of phenomena, e.g., some bear a negative charge and some a positive charge. But even if all of the smallest particles were of one sort, this would not preclude certain arrangements of some of them from resulting in different kinds of items.

---

[21] Cf. below, n. 25, p 155.
[22] Cf. below, pp. 149-50 and 155 ff.
[23] See Smart, "Sensations", p. 161 and Feyerabend, "Materialism", p. 49.

Sensations are good examples of such items. And if theories (and words) are composed of elementary particles, they too are good examples of phenomena which cannot be reduced to their material components.

Feyerabend says [24] that the sound we hear may appear to be different from its mechanical causes, or 'manifestations' as he calls them, but is really nothing but these causes or manifestations. His claim that the seen table appears to be something different from the felt table is not at all enlightening. Nobody is ever surprised that the table which he sees is one and the same as the table which he feels. We would expect our different senses to reveal the presence of an object in different ways. It is paradoxical, however, to claim that *heard* sound is one and the same thing as its mechanical manifestations. These mechanical manifestations, after all, were discovered by our seeing and hearing things (by our experiences). If we are constantly deceived in thinking that heard sounds are heard sounds and not something which is *really* mechanical, we could scarcely rely on our senses when investigating the causes of sensation.

Ernest Nagel criticizes the sort of reductionism in which Identity theorists indulge. In "The Meaning of Reduction in the Natural Sciences", he says the following:

....if and when the detailed physical, chemical, and physiological *conditions* for the occurrence of headaches are ascertained, headaches will not thereby be shown to be *non-existent* or *illusory*. On the contrary....all that will have happened is that the occurrence of headaches will have been *explained*.[25]

It is a mistake simply to identity all phenomena with what causes, constitutes or accounts for them. Something cannot cause or be a constituent of itself; it must cause or constitute something else. To find the causal conditions and/or the constituents of X and then to solemnly declare that X is nothing *really* but its causal conditions or constituents, is to contradict oneself. Yet if Identity theorists are not explicitly denying that we have sensations as ordinarily understood, they must be interpreted as saying that once we know all the neural correlates of pain or any other sensation, we will realize what the true nature of pain is; we will then know pain for what it *really* is.

It may be objected that in some cases we can identify something with what causes it. For example, the movement of the words flashing the

---

[24] See above, p. 78.

[25] In *Readings in Philosophy of Science*, ed. by P.P. Wiener, p. 549, emphasis mine.

news across the Times Square building is caused by lights going on and off. Therefore, the movement of the words across the building is *nothing but* the lights going on and off.

I agree that some things may be identified with their causal conditions but the expression 'nothing but' requires examining. Just as a storm may be the sum total of its causal conditions or constituents, so the movement of the words may be constituted by the lights flashing on and off. And the expressions 'the storm' or 'the movement' do not refer to any ghostly entities or processes over and above the causal conditions, no more than 'sensations' refers to ghostly entities over and above the neural processes. However, what is implied by the expression 'nothing but' in the above statement? Surely the lights cannot be flashing in a haphazard fashion. They must be flashing in a particular sequence to produce the movement of words. Thus the movement of words is constituted by the flashing of certain lights in a particular sequence. It is not false to say that the movement of the words is nothing but the lights going on and off if this is not intended as a denial that words do seem to be moving across the building. But the whole point of insisting on the 'nothing but' seems to be that less is going on than we might think.

In this particular case it might be suggested that there really is no movement of words but only particular lights going on and off in certain sequences. This is what a reductionist would claim. But one who makes *this* claim cannot say that the movement of the words is caused or constituted by the lights going on and off. He must say that the illusion of the movement or the apparent movement is caused by the lights. But if the reductionist's claim were true in this case, it would not mean that it was true in all cases. The example chosen is such that it is plausible to say that the phenomenon in question, i.e., the movement, is an illusion. A comparable example might be the claim that what we see on a cinema screen is really nothing but the extremely rapid flashing of slightly changing still pictures.

I do not wish to defend the view that all phenomena are irreducible to their causal conditions. It may well be the case that flying saucers, the movement of words produced by lights going on and off, and the movements on a cinema screen are all reducible to their causal conditions or constituents. However, in the case of phenomena which are not illusions, such as pains or words, it is entirely misleading to say that they are nothing but their causal conditions or constituents. The point is that simply to identify pains with their constituents is to imply that pains are somehow illusory.

This does seem to be implied by Identity theorists. It is as though investigation into the causes of a disease are conducted and then, when all its causes are discovered, the pronouncement is made that the disease is in *some* sense an illusion, that it is *really* certain germs acting on the tissues of a certain organ resulting in such and such symptoms. Physicians do not make such a claim. But a philosopher might interpret certain statements made by physicians as supporting the view that diseases are "nothing really but", or are "identical with", their causes and symptoms. For example, a physician may say that diabetes is a lack of insulin resulting in an excessive discharge of urine. If one were to deny that this is what diabetes is, both the physician and the philosopher might suspect that perhaps the denier was postulating something like possession by 'diabetes demons' to account for the disease. But one need not be sympathetic to witch doctors to question reductionism. It is the reality of diseases, not of demons, that is in question here.

When a physician says that diabetes is a lack of insulin resulting in certain symptoms, he is not denying the reality of diabetes. On the other hand, when an Identity theorist says that sensations are nothing but brain processes, he implies that pains, for example, are not what we have always thought them to be, viz., a characteristically distressing experience. He may admit that people do have experiences and he may be right to point out that they are not ghostly processes but are brain processes. However, his insistence that they are *nothing but* brain processes and his suggestion that we could cease talking of 'pains' at all, seem to go beyond anything that the physician says in connection with diseases. So far as I know, no physician has advocated replacing talk about diabetes (the whole disease) with talk about insulin deprivation and excessive urination (the factors in the disease). If a physician did advocate this, we would wonder whether he was implying that we had been mistaken all along in thinking that there was such a disease. For him to advocate cause-and-symptom-talk instead of diabetes-talk would lead us to suspect that he refused to recognize the total disease we call 'diabetes'.

Whenever we find the advocacy of talk in terms of parts instead of in terms of the whole, we can only conclude that the reality or existence of the whole is being called into question. I do not object in principle to the denial of the existence of certain phenomena but to deny the existence of diseases or pains because we have discovered their causal conditions and/or constituents commits us to the denial of *everything* which we come to understand. It would be as if we learned all the causal conditions of pollution and then advocated dropping talk about 'pollution' in favor of talk about excessive mercury or carbon monoxide in the water or air and the effects thereof.

If the Identity Theory comes to the view that there are pains as normally understood but these are *really* nothing but brain processes, then I find the Identity Theory to be incoherent. This is because the brain processes are constituents of a whole—the experience of pain—and to deny the existence of the whole commits one to a denial of its constituents. As normally understood, pains are distressing experiences; they are objectionable because of the way that they *feel* to the individual. If we have learned that they are constituted by brain processes and not ghostly processes, this does not mean that they are illusory or nonexistent. The incoherence of the Identity Theory becomes most evident when Identity theorists affirm the existence of pains yet insist that they are "nothing but" brain processes. (Compare to: "The movement of the words across the building is nothing but....")

This incoherence can be shown in other ways. For example, it has been argued here that there is an asymmetry, a difference in kind between the way in which I know that I am in pain and the way in which another knows this. I do not need to observe or infer anything in order to learn that I am in pain. This fact, among others, is what makes us call pains 'mental' processes. This asymmetry in the way we learn of the sensations of others and are aware of our own must be taken into account by any philosophical theory of the relationship between sensations and neural processes. Even if we decided that the invariable correlation between sensations and neural processes amounted to an identity, we would *still* have 'mental' processes. We would, that is, still know certain sorts of things about ourselves in a radically different way from that of others. Even if we deny the existence of all but physical processes, as Smart would have us do, we would still have certain processes (whether we call them physical or mental or physical *) which would be differentiated from all other processes because of the radical difference in the way in which they are discovered by others and by the person undergoing them.

It may be objected at this point, but not by an Identity theorist, that the reason for this asymmetry is the fact that the processes in question are *mental* ones. It may be said, that is, that we know what is in our minds in a different way from that of others because it is *in our minds*, because it is somehow *mental*. I am arguing that such processes (or states) as pains are called 'mental' because, in part, of the asymmetry between how others know one is in pain and how the one in pain knows it. The mind is not a ghost as Ryle and Smart say it is usually conceived. Mental processes such as sensations are not ghostly. They are processes or aspects of processes whose existence is determined in ways totally unlike objects or processes which are intersubjectively observable. A person's non-inferential knowledge of his own pains (and other sensations) has led some philosophers to call pains 'private' and to call their owner's relationship to such pains a 'privileged' access. But this is not tantamount to postulating a ghostly realm. It is not in the least surprising that the person who has the pain should be in a different epistemological relationship to it than others are. He will have a different basis for saying that he is in pain than others will. As Wisdom says, in *Other Minds*, "....to know one is in pain is to say on the basis of pain that one is in pain".[26] But to say that someone else is in pain is not to say it on the basis of pain that one feels oneself. One says it on the basis of the fact that the one in pain exhibits pain behavior and/or that he informs one that he is in pain. Or, once neurophysiology has become perfected, one could perhaps say it on the basis of observing the appropriate neural processes (or recording them on a perfected EEG).

It may be asked how I account for the difference in the epistemological relationship which one has to one's own pain and another's pain, if not in terms of a ghostly mind. In addition to what has been said immediately above, viz., that the one in pain *has* the pain and others have to make same sort of observation of him in order to learn of his pain, I would add the following.

The epistemological disparity arises not because we have ghostly minds but because our bodies, including our nervous systems, are independent of each other.[27] If I do not suppress the natural signs of my pain and/or if I tell you about my pain, you can know as well as I do that I am in pain. The epistemological disparity upon which I am insisting refers more to our *way of telling* that a person is in pain than to our knowledge of his pains. Sometimes we can hide our pain, sometimes we cannot hide it and/or do not wish to hide it. The point is that our pain could not be hidden from us though it can, on occasion, be kept from others. But even if neurophysiology were perfected and others could know that we were in pain despite our efforts to hide it, we might still know better

---

[26] *Op. cit.*, p. 173.
[27] Cf. below, pp. 136-37 and p. 139.

than they do, at least some (if not *all*) of the time, exactly how our pain feels to us—particularly if the same neural processes could constitute different sensations, on occasion.

It has been claimed that if neurophysiology became perfected in the sense that we could always tell whether someone were in pain by observing his neural processes, this would show that persons in pain do not have a privileged access to their sensations in the sense that their sensation reports are incorrigible. There may come a time, so the argument goes, when a person will no longer be considered the ultimate authority regarding his own sensations. Neurophysiological theory may advance so far that if someone claims to be in pain but is not undergoing the appropriate neural processes, then we shall refuse to accept his claim to be in pain. In such a case, it is argued, we would refuse to abandon our strongly corroborated empirical correlations between sensations and neural processes and would say that the individual was somehow mistaken or prevaricating. Smart considers such a possibility on page 99 of *Philosophy and Scientific Realism* and Armstrong considers it on pages 107-108 in *A Materialist Theory of the Mind*. I think that an examination of one of the assumptions underlying such a possibility will reveal once again the incoherence of the Identity Theory.

Armstrong argues that we do not have a logically privileged access to our own mental states, that we are not the "logically ultimate authorities on our inner states".[28] This is how he argues for this view:

Once it has been admitted that I can be wrong about my current inner states, then we must allow the possibility that somebody else reaches a true belief about my inner state when I reach a false one. Now might not this person have good *reason* to think that a mistake had occurred? Suppose that certain sorts of neurological processes were necessary for the occurrence of pain. Suppose a person reported that he was in pain, but in fact he was not in pain, and that an observer discovered that the requisite brain-process had not occurred. If brain theory were in a sufficiently developed state, might not the observer conclude with good reason that the subject were not in pain? It might be objected that the observer would have no way of ruling out two other hypotheses: (i) the subject had made an insincere report; (ii) the brain-theory previously developed had been falsified. Now no doubt the logical possibility of these hypotheses could never be ruled out, but if enough were known about the behavioural and physiological correlates of mental states might not these hypotheses be ruled out for all practical purposes? And, if so, the observer would be a better authority than the subject on the subject's mental state.[29]

On the next page he invites us to "Consider the case of a brain technician

----

[28] *Op. cit.,* p. 107.
[29] *Ibid.,* p. 108.

who has a perfect understanding of the correlation between the states of my brain and my mental states."

The assumption which Armstrong is making here is that better evidence, or a better criterion, of someone's being in pain than his own report might very well be the occurrence of certain neural processes which are known to be correlated with the occurrence of a mental state such as pain. Such a criterion could, according to Armstrong, supercede even a person's sincere avowals.

I am not here concerned to dispute Armstrong's argument against a logically privileged access. He may, indeed, be right in the sense that if neurophysiological theory were perfected, then we could tell as surely as the individual himself whether he was in pain or not. What I am concerned to show here is that the assumption underlying this argument used by Smart and Armstrong is incompatible with the claim that sensations are identical with, are nothing but, neural processes.

It might be said that since the one in pain has a different basis for saying that he is in pain than the neurophysiologist does, the former must know with more certainty that he is in pain than the latter does. It is true that the one in pain will always have a *different* basis than do others, viz., his *having* the pain, but in the case under discussion, the neurophysiologist might have a basis which is just as good. Given that our correlations *are* perfected, then others could know that we are in pain just as surely as we do. They could observe the neural processes which invariably occur simultaneously with, and perhaps constitute, pain. But even if neurophysiology is never perfected, others can know as well as we do *that* we are in pain. If we suddenly stab someone with a knife and he immediately gives every indication that he is in pain, we know as well as he does that it hurts, i.e., that he is in pain.

The disparity between my knowledge of my own pains and of yours shows itself most clearly in cases where I successfully conceal my pain from you or successfully convince you that I am in pain when I am not. On the other hand, if *I* am convinced that I am in pain, then I cannot be mistaken; nor could my own pain be concealed from *me*. On most occasions we know as well as the one in pain that he is in pain but there are times when he can hide it or feign it. Also, there are occasions when we may have trouble conveying to another exactly how our pains feel to us, but these cases are relatively rare.

If the occurrence of certain neural processes could constitute incontrovertible evidence, or a decisive criterion, that someone were in pain, which is what Smart and Armstrong suggest, then these neural processes could not themselves be the experience or sensation of pain. According to the dictionary, a criterion is a canon or standard by which anything is judged or estimated. Or, it is a characteristic attaching to a thing, by which it can be judged or estimated, i.e., it is a *part* of that thing.

According to Armstrong and Smart, in the case at hand we judge that someone is in pain, is in a certain 'mental' state, on the basis of the fact that he is undergoing certain neural processes. We can tell by the neural processes that he must be in pain. Our well established and detailed *correlations* enable us to use the neural processes as the criterion of the person's being in pain. This is Smart's and Armstrong's reasoning and, so far as I can see, it is perfectly intelligible.

However, its intelligibility shows again the incoherence of the Identity Theory. How could sensations be identical with, nothing but, or "really" neural processes if the latter could be the criterion of their occurrence? If neural processes could be a *criterion* of pain, how could they simply *be* pain? If N is the criterion of P, then it cannot be (identical with) P. A criterion cannot be a criterion of itself; it must be a criterion of something else—even if that something else be the whole of which it is but a part. This shows conclusively, I think, that sensations cannot be (identical with, really, nothing but) neural processes. They are processes which are known to the individual undergoing them in a completely different way from that of other persons. The one in pain does not need, and could not sensibly be said to use, a *criterion* to determine whether he is in pain. His knowing that he is in pain is necessitated by his having the pain. Others do need a criterion to determine whether he is in pain. This shows that pains are not *just* neural processes even if the latter were their sole criterion.

It may be argued that we sometimes *do* identify S with B, or say that S is really B, where B is a criterion of S. For example, if a patient complains of a pain in his chest, he might be suffering from indigestion, a heart attack or any of several other conditions. When the physician determines that it is a heart attack, might not his criteria for determining this also be considered one with what a heart attack really is?

Let us suppose that the physician bases his judgment on a combination of clues and symptoms which points unmistakably to a heart attack. This combination of symptoms and clues is the criterion of a heart attack, e.g., a chest pain, an increased pulse beat, a numbness in the left arm, difficulty in breathing, obesity and a history of heavy smoking. Once we have eliminated the possibility of indigestion, etc., by means of these symptoms and clues, we have identified the condition, we know what it really is, viz., a heart attack. Are the criteria one with what the heart attack really is? (Is a discharge of urine one with what diabetes *really* is?) [30]

We must answer "No" to this question. Even if certain symptoms of a disease count as the criterion of that disease, these are not *themselves*, say, the heart attack or what a heart attack *really* is. The effects and the causes of a heart attack may be deemed parts of it, but the attack itself is the stretching or bursting of the walls of the heart or the irregularity of the valves of the heart caused by certain micro-organisms, such as the pneumococcus, etc. Were the patient *transparent* and the physician somehow able to observe the attack itself, would the physician require criteria in order to determine what was occurring? Would he be likely to say that the stretching or bursting of the walls or the lesion taking place in the valves was a symptom or criterion of a heart attack? I submit that if he did say something like this, he would be using the expression 'heart attack' to refer to the *whole* condition in question, i.e., to the events which caused it, the effects which follow it, and the attack itself.

If he were witness to the actual attack, he would not likely call what he observed an indication, a symptom or a criterion of a heart attack. He would say that he was witnessing an actual heart attack. And if he could observe a patient's femur breaking, he would not be likely to call this the criterion of his patient's femur being broken.

The signs, symptoms, indications and criteria of a phenomenon are neither simply identical with nor are 'really' that phenomenon itself, though they may be a part of the whole phenomenon. In other words, items are no more reducible to their criteria than they are to their parts. The causal conditions and criteria of an item—unless the phenomenon is an illusion—may be considered to be a *part* of that item but this is not to say that they simply *are* that item or that they are what it *really* is.

If we distinguish between '*one with* what the heart attack really is' and 'what the heart attack *really* is', we can say that the criteria may be a part of, and in this sense 'one with', the attack itself but they cannot be the whole condition or what the attack *really* is. A heart attack is a heart attack, not the criterion of a heart attack. A pain is a pain, not the criterion of a pain.

It should also be noted that Armstrong speaks of correlations between mental and brain states in the above quotation. As has been mentioned before, it is, to say the least, difficult to see how we could correlate

[30] Cf. above, p. 87.

something with itself. If mental states are simply brain states, as Armstrong argues, then it would seem to be impossible to establish empirical correlations "between" them (it).

Perhaps it will be objected here that N can be a sign or evidence or criterion of P and yet at the same time *be* P. For example, approaching clouds may be both a sign or criterion of an imminent storm and they may also be part of that storm when it does arrive. However, though they may be part of the storm, they could not constitute the entire storm. If we saw the clouds and the rain approaching, we could not claim to have a criterion, or to see the evidence or signs of, an imminent storm. We would see the storm itself. If we were looking for a missing loaf of bread and found some crumbs from it on the ground, we might say that we had found signs or evidence of the missing loaf. But if we found the loaf itself, we would not say that we had found further signs, clues or evidence of the loaf. We would have found the loaf *itself*.

I realize that we do not speak of 'criteria' of loaves of bread and that there are distinctions to be made between signs, evidence and criteria. What I am arguing is that anything which constitutes our *way of telling* that X is the case cannot simply be identified with X. When we know that we are confronted with X itself, there is no need for recourse to a way of telling whether it is X or something else. The need to use a way of telling about X arises only when we are not confronting X itself or are trying to ascertain whether it is X which we are confronting. If sensations were nothing but brain processes and we could observe the brain processes, the question of whether the individual were in pain or not could not arise. His being in pain would just *be* the occurrence of these brain processes. Nor could the question whether X is having a heart attack arise if we were actually observing his heart attack. In general, if S is nothing but B, then if we were given B, we would *ipso facto* be given S. Where there is a possibility of doubt, as in the sensation-brain process case, this is proof that the one cannot simply be the other.

Neither a storm, a loaf of bread, a pain, nor anything else can be a criterion, evidence or sign of *itself*. A criterion is our way of telling about what it is a criterion *of*. The same applies to signs and evidence. It may be that N can be a sign or evidence or criterion of P and still be part of P, e.g., the characteristic signs of anger may be part of anger, but N could not be the whole of P.

Perhaps it will be suggested that the signs of anger may be all that "really matters". Leaving aside the question of *to whom* these may be the sole concern, I would say the following. Surely a violent display of anger and the angered person's feelings and/or reasons for his anger are no less important than the signs of anger, e.g., "A twitch of the eyebrow, pallor, a tremor in the voice....".[31] It is difficult to say in general what will matter to those involved in cases of anger but we can say that the signs of anger are unlikely to be the sole or primary concern of everyone involved. Lastly, if the suggestion amounts to the claim that anger may be nothing *really* but its signs, then we can reject it on the grounds that our way of telling about something cannot be what that item *really* is. A person might well be angry when alone and it is difficult to see how a plausible case could be made for the signs here being "all that matter".

Any physiological sign or criterion of pain could not (logically could not) constitute the whole of pain. Materialists cannot have it both ways —they cannot identify pains with neural processes and also hold that the latter could be a criterion of the former. The Identity Theory now seems to be not "S = B and S is really just B" but "S = B, S is really just B, and B could be the criterion of S". It would seem that if the Identity Theory cannot be shown to be false, this may be because it is unintelligible. It is unintelligible because it amounts to the claim that "Sensations are really brain processes; the latter could very well be an unmistakable criterion of sensations". Since it is incoherent, the most one can do is to show this incoherence and the inevitable obscurity of any so-called theory which, in effect, denies the existence of what it purports to explain.

Identity theorists do not intend their identification of sensations with brain processes to be a denial of the occurrence of sensations. They know as well as anyone that sensations occur. What they claim to have discovered, or predict will soon be discovered or decided is, as we have seen, that sensations are really brain processes. Just as the Morning Star was mistakenly thought to be something different from the Evening Star, so sensations have been erroneously thought to be different from brain processes. In fact, they argue, though the connotations of 'Morning Star' and 'Evening Star' are, and will continue to be, different, the denotation of these two expressions is one and the same, viz., Venus. In the same way, though the connotations of 'pain' and, e.g., 'C-fibers are firing' are different, their denotation is one and the same, viz., a process in the brain.

To be fully acquainted with, to know everything about, the Morning Star (Venus) is to be fully acquainted with the Evening Star (Venus).

---

[31] J.L. Austin, "Other Minds", in *Philosophical Papers,* p. 75.

Is this true of brain processes and sensations? Does knowing all about the former entail knowing all about the latter? I think not.

I do not think that the sense-referent distinction is of any help to Identity theorists. I think it can be shown that someone who has never had the sensation of pain will not understand what pain is, even though he may understand what neural processes are and know all about them. He could be familiar with the referent of 'neural process', even familiar with the neural processes allegedly identical with pains, and yet be totally ignorant of what pain is (i.e., of the referent of 'pain'). If this is the case, as I believe it is, then doubt is cast upon the serviceability of the sense-reference distinction to Identity theorists. In fact, it casts doubt upon any attempt to identify sensations with anything else or to analyze them in terms of anything else.

We can imagine visitors coming to earth from a distant planet. We can imagine that these visitors (call them Martians) are so constituted that they have never experienced, and could never experience, pain. We may suppose that we can communicate with the Martians about all the usual sorts of things but that they are utterly perplexed by our pain behavior (avoidance and remedial) and by our references to pain. How would we go about helping them understand what pain is? Presumably for the Identity theorist there would be no problem. He would, we may suppose, explain to the Martians that there is really no problem here at all. He would explain that since the Martians have already learned how we use 'pain' and 'neural processes', all that remains is to familiarize them with the neural processes (the common referent of 'pain' and 'neural process') identical with pain and nothing more need be done. Since, according to the Identity theorists, we have here two different senses (or connotations) but only one referent (the neural process), the Martians should now know all that we do about pains (i.e., neural processes). Once they know what the appropriate neural processes are, they will know what pains really are. If sensations are strictly identical with certain neural processes, as Smart says they are, then surely to know what the appropriate neural processes are will be to know what pains are. And in case the Martians still profess ignorance about the nature of pains, then presumably Feyerabend would tell them that "....if you want to find out whether there *are* pains, thoughts, feelings in the sense indicated by the common usage of these words, then you must become....a materialist".[32] I would suggest that if the Martians wanted to find out what pains

---

[32] Feyerabend, "Materialism", p. 53.

are, what 'pain' refers to, they would have to become acquainted on a first-hand basis with pain. They would have to experience pain if they wanted to understand the referent of 'pain' as we do.

It may be argued that if the Martians were to be acquainted with the appropriate neural processes, in the sense of *undergoing* them, then they would be acquainted with pain. Since sensations *are* brain processes, it will be said, it is not surprising that the Martians do not know pain as we do. To experience the brain process *is* to experience pain—only those who undergo the appropriate brain process will experience the pain.

This argument does not prove that the Identity Theory is true. Its conclusion is compatible with any Dualism which admits a correlation or correspondence between sensations and brain processes. The Dualist could argue that one must undergo the appropriate brain processes in order to experience pain because they cause, are caused by, are invariably correlated with, or are a feature of the state of pain. The Dualist could argue that the awareness of pain and the brain process are both aspects of being in pain, the former being the subjective aspect (the contingently "private" aspect) and the latter being the objective aspect which is in principle intersubjectively observable.

Thus the fact, if it is a fact, that the Martian has only to undergo the appropriate brain process in order to undergo pain establishes an identity no more than it establishes a causal interaction or an invariable correspondence. The Identity Theory is not so much an independently testable scientific theory as it is a philosophical attempt to reduce the number of entities in the world.

However, it is misleading to suggest that the word 'pain' refers to an entity (thing) of any sort, including a process in the brain. To be in pain is to undergo a kind of experience. This experience may be mildly annoying, rather vexatious, very distressing or utterly unbearable. Though every experience of pain may involve certain neural processes and every degree of pain may involve a modification in these neural processes, one who knew only about these physiological processes would not thereby know about the experiential aspect of pain.

The Martian might know that the brain process associated with painful experiences, or the experience itself, produces a spike pattern on an EEG and that the more distressing the pain is, the higher the spikes appear on the chart. But the sort of variations he will detect between the neural processes associated with a mildly annoying pain and those associated with an excruciating pain are of a different sort than the variations between the experiences themselves. That is, the difference between a

mild pain and an excruciating one is a different *sort* of difference than
that between, e.g., a slow and infrequent firing of certain neurons versus
a rapid and continual firing. The former is a difference in how the pain
feels to us; the latter is a difference in the speed and frequency of the
occurrence of certain processes in the brain.

I do not think, then, that the Identity theorist is better placed than
others to acquaint the Martians with the referent of 'pain'. If he neglects
to mention the experiential (subjective) aspect of pain, he will be omitting
its most important feature.[33] To say that 'pain' has no referent but the
neural process would only confuse the Martians. The Identity theorist
would find himself facing this question : "We can understand what the
referent of 'C-fibers firing' is but what is the referent of 'pain' "? In
short, they still will not understand what all the shouting (moaning,
groaning, shrieking or writhing), i.e., pain-behavior is about.

In *Philosophy and Scientific Realism,* Smart says of a congenitally
blind man that he could know the meaning of 'red', 'green' and 'blue'
just as well as the rest of us.[34] Presumably then, Smart would say that
the Martians could know the meaning of 'pain', even though they had
never felt pain. But neither the congenitally blind man nor a Martian is
acquainted with the denotation (referent) of the words in question, though
both may be conversant with the concepts and connotations in question.
The Martians may know the rules for the use of 'pain' and will eventually
be able to recognize cases of pain from the behavior of others but this
does not mean that they know the referent as the rest of us do. If we
distinguish between what a word refers to and what it means (connotes),
then we can agree with Smart that the blind man and the Martians
can know what 'red' and 'pain' mean but disagree that they will know
what the words refer to as we know it.

Pains cannot be identified with anything else nor can they be completely analyzed in terms of anything else. Telling a congenitally blind
man that a certain color is identical with a certain wave-length of light
would not acquaint him with the referent of color. Telling a Martian
that pains are identical with the neural processes which he was observing
in someone would not acquaint him with the referent of 'pain'. One
would have to be an appropriately equipped organism to know at first
hand what the referents of 'red' and 'pain' are.

---

[33] It is most important in the sense that pain-behavior could not be understood by
the Martians without taking it into account.

[34] *Op. cit.,* p. 75.

CHAPTER FIVE

THE PHYSICAL AND THE MENTAL

Different kinds of things can be said of experiences than can be said of brain processes. This indicates, *prima facie,* that these are different sorts of phenomena or at least that they cannot simply be identified. Because we can call pain "excruciating" or "agonizing" and cannot sensibly make such ascriptions of brain processes,[1] we know that pain cannot be simply a process in the brain. Similarly, we can sensibly locate a neural process in a particular place in the brain but not an experience. Since organisms have experiences, we can say that their experiences occur wherever they (the organisms) are, but this does not mean that we can locate them in a particular part of the body (including the brain); no more than we can locate concepts or ideas in a particular part of the body.

If one says that experiences occur in our heads or brains, this is shown to be nonsense when the question is asked whether it occurred three inches behind the left eye or perhaps more toward the right side of the head or brain. How would we go about answering this question? It is incapable of being answered sensibly because it is incapable of being asked sensibly. It makes no more sense than does the question, "What (or, Where) is the square root of green?".

Experiences cannot be assigned a bodily location—they are not the sort of thing that can be so located. Bodily location is not a possible property of an experience. Sensations, e.g., pains, can be given a location but they are not located in the same sense in which cuts or bruises or our limbs are located. Nobody could put his finger on my experience of

---

[1] Two scientists are observing a subject's brain processes. It is obvious which of the following remarks are about the brain or its processes and which are about the experiences associated with them: "Look at the pattern they make"; "They must be excruciating"; "Observe this one over here"; "Place the probe between them"; "This one is four inches to the right of the former one".

pain as he could put it on my injury, nor could a surgeon find my experience of pain during an operation as he could find the source of my pain. We say we have a pain 'in' our arm or leg, not 'on' it. But it is not in our arm in the same sense in which our muscles or bones are in our arm.

Neurons, on the other hand, are in our bodies in the sense in which muscles and bones are. They are located in space. Microphotographs have been taken of neurons but none has been taken of experiences, e.g., of pains. The reason for this is not lack of equipment. It is because pains are not the sort of thing which could be photographed. This shows that neurons or neural processes have a property which pains do not have, viz., a particular bodily location. This shows that pains could not be identical with neural processes.

If pains, or the having of pains, or the experiences or sensations of pain, were "strictly identical" with neural processes, as Identity theorists say they are, then they would have every property which neural processes have. If they lack a property which neural processes have, or have one which the latter lack, then they cannot simply be identified with neural processes. If they could be identical (one and the same thing) yet have different properties, then this is a new sense of 'identity' and the Identity Theory means something different from what it purports to mean.

It may be objected that just because different things can be said of sensations or experiences than can be said of brain processes, this does not in itself mean that these cannot be one and the same item. For example, at one and the same time a person may be described or referred to in any and all of the following different ways: 'Joe's father', 'Bill's son', 'the tallest man in the room', 'the mayor of Thunder Bay', 'my best friend' and 'him'. Here we have one item with many different labels or descriptions and perhaps even different *sorts* of labels or descriptions. Why, the objection continues, may not 'sensation' (or 'pain') and 'brain process' merely be two different ways of describing or referring to the same item, viz., what someone feels (sensation) and what another sees (brain processes).[2]

The mere fact that 'Mentalese' cannot be translated without loss into 'Materialese', it is concluded, does not in itself mean that what someone feels as a sensation is not the same item as the neural firings that another can, in principle, see. After all, expressions like 'Bill's son' mean some-

---

[2] See, for example, Smart's "Sensations", pp. 168-69.

thing different—convey different information—from expressions like
'the   tallest man in the room', yet only one referent may be involved,
i.e., Bill's son may *be* the tallest man in the room.

Thus an Identity theorist may conclude that I have failed to show
that sensations and brain processes are two different things. And I would
be the first to agree with him, depending on what is meant by 'two
different things'. This is because of the ambiguity of such expressions
as 'the same item' or 'two different things' in this case.

It is not being suggested that we are always or most of the time unclear
about the meaning of these expressions. Most of the time there is no call
for doubt or confusion. However, in the case at hand, I would argue
that we have a problem. Take for example the following question: "Is
what A feels as a sensation and what we see as brain processes one
and the same item?" Our answer here should not be a straightforward
"Yes" or "No" but must be "Yes *and* no". In the following, I will
show why this is the case.

It is crucial to realize that while it *is* sensible to describe one person
in any and all of the above ways, it does seem to be meaningless—to
convey no information—to describe experiences as 'circular'[3] or brain
processes as 'exquisite'.[4] Where we can say *sorts* of things about S that
cannot meaningfully be said of B, and vice versa, this is at least a *prima
facie* indication that S and B are not simply 'one and the same' items.
But this need not mean that they are *entirely different* items. My view
is that in a sense they *are* the same item but in a more important sense
they are *not* simply the same item. Is the word 'form' the same item
as its constituents? In a sense "Yes" and in a sense "No".

It is my contention that we cannot simply identify sensations and
brain processes. But this does not mean that these are two different
items. The apparent paradox—the same, yet different—is explained when
we realize that to restrict our answer to either 'the same' or 'different' is
to mask crucial distinctions. A thing may be identical with its constituents
in one sense but not identical with them in another sense. Thus the
Identity Theory may be true, but it is as trivially true as an Identity
Theory of words which informed us that words are (nothing but) letters
(or sounds).

My view is that S may *be* B without being *reducible* to B. Sensations
may be constituted by neural firings, just as words may be constituted by

---

[3] Cf. below. p. 107.
[4] Cf. below, p. 150.

letters or sounds, yet it is highly misleading to say that sensations (words) are *merely* neural firings (letters or sounds). If asked to describe *what is occurring* in one who is in pain, it would be misleading to say that there is nothing occurring but neural firings. The question "Who is that?" might be sufficiently answered by "The mayor of Thunder Bay" or "Bill Anderson" or "My best friend", depending on the context, but a Martian who asked "What is pain?" would not be sufficiently answered by "Pains are nothing but neural firings". (The Martian would likely ask why the words "nothing but" are included in the definition.)

One could say that sensations are brain processes but to think that this is the only important thing to say about sensations is to ignore the most crucial and characteristic feature of being in pain, viz., the fact that it is an experience. One might just as well claim that the *real nature* of human beings, the *final* word about them, is that they are bones, blood, skin, etc. I am not denying that persons are bodies. I am suggesting that to claim that persons are 'nothing but' or 'really' or 'simply' bodies is to consider the *first* word about persons to be the *last* word about them. Persons after all, are not just objects (bodies) but are subjects too —something that Smart, for one, would like to deny.[5]

Perhaps some analogies will be helpful here. Suppose it is insisted by a reductionist that a watch is nothing really but two hands, a face, numbers, a mainspring, etc. A watch, he argues, is one and the same item as its constituents. What is the proper response to such a claim? I submit that it would be the same argument as that which I have been presenting in regard to sensations and brain processes. And one way in which our case could be argued is as follows.

It makes sense to say that one can tell the time by a watch but it does not make sense to say that one can tell the time by two hands, a face, a mainspring, etc., all occurring randomly. Now, it is not the case that a watch is a different item than its parts which are two hands, a face, a mainspring, etc. But it is also not the case that a watch is *nothing but* (and in this sense identical with, or the same as) two hands, a face, etc. A watch is all its parts arranged in such and such a way and functioning in such and such a way. One can tell the time by a watch, by what is composed of these parts, but it does not make sense to say "one can tell the time by a mainspring, a watchstrap, etc., merely lying in a heap".

Thus, in a sense a watch is nothing but its parts. But in a more important sense it is not simply its parts. The analogy with sensations and

[5] Cf. above, p. 14.

brain processes is not, of course, a perfect one, nor is the analogy of words and letters a perfect one. For example, sensations cannot be said to have parts in exactly the same sense as watches can, and only in *some* cases of words is it sensible to speak of 'parts'.[6] All of these cases are intended to show that even if S is composed of or constituted by B, even if in this sense it *is* B or is *identical with* B, it may still not be *nothing but* B. It may still not be *reducible* to B; it may still not *simply* be identified with B. And one way of showing this is to show that some things which can sensibly be said of B cannot sensibly be said of S. Thus the neural firings which constitute pain may occur in a certain part of the brain but it seems senseless to locate the experience itself in space.

Is a box nothing but what it is composed of? The question is ambiguous. This is why our answer must be ambiguous too. We must say "In a sense yes and in a sense no". And we could go on to argue that it makes sense to say that one is sitting in a box but not that one is sitting in the sides of a box or in a heap of cellulose fibers or a heap of molecules or atoms or electrons.

What we are maintaining is that there are *different kinds* of things in the world, including sensations, whereas Identity theorists and reductionists in general think that there is 'basically' or 'really' just one sort of item, i.e., one which is composed of, and so reducible to *nothing else*.[7]

Is a piece of kleenex nothing but a heap of a certain type of molecules? If it makes sense to say that one can blow one's nose with a piece of kleenex, but not with a mere collection of these molecules or atoms, then there must be some sense in which pieces of kleenex are not reducible to their constituents or to their 'ultimate' constituents. And the same applies to sensations and brain processes. The point is that we can grant everything to the Identity theorists yet reject their conclusion, i.e., that sensations are nothing but brain processes.

Given that it makes sense to say that one can tell time by a watch, blow one's nose on a piece of kleenex, hide in a box, or that a pain is excruciating and that it does not always *make sense* to say such things about the constituents of these items, then these items can be legitimately regarded as irreducible to their constituents. The mere fact that we talk

---

[6] Could the word 'a' in "I see a radio" be said to have parts? Could the word 'radio'? (Would the parts be the five letters or the three syllables? Both?) This shows us that words may, in a sense, *be* letters, but in a more important sense they are *not simply* letters.

[7] Or nothing? Cf. below, p. 157.

of 'constituents' here—or 'components' or 'parts'—indicates that there must be something of which they are constituents.

It must be emphasized that though S is composed of nothing but B's, S is not *ipso facto* "nothing but", or is not reducible to, B's. This applies to any S or any B.

For example, it may be argued that a heap of marbles is comprised of nothing but marbles and that therefore it *is* nothing but marbles, or is identical with these marbles. A heap of marbles, in other words, is *nothing but* or just *is* marbles. If this is true, then it may be true of *heaps* of anything but it is not true of words, sensations, watches or boxes. (And what information *would* be conveyed by the claim that a heap of marbles is 'really' marbles?)

A heap is just many things lying together in a pile. It is a mass or haphazard collection of items. However, in the case of a word, a watch or a sensation, the constituents are related and/or interact in such a way that a different sort of item results. This is indicated, I have argued, by the fact that different *sorts* of things can be said of words or sensations or watches or boxes than can be said of their constituents, either individually or as a mere collection or heap.

We may concede everything to the Identity theorists, therefore, in regard to what comprises sensations. Just as "a cloud is a mass of tiny particles" [8] and just as lightning "is a motion of electric charges",[9] so sensations are brain processes. But in all these cases, the equation of identity is reversible. Thus a mass of a certain type of tiny particles constitutes a cloud; a certain motion of electric charges constitutes lightning; certain neural firings constitute a sensation. But we have argued that in the case of at least some phenomena, e.g., words, though they *are* their constituents (words *are* letters), it is highly misleading to speak of an identity here. Words are not simply *heaps* of letters, nor are sensations simply random neural firings. It is as unenlightening, as fruitless, to characterize sensations as being simply (or 'nothing but' or 'really') neural firings as it is to characterize words as being nothing but or nothing really but letters. (I ignore spoken words here.)

It is not so much what Identity theorists assert ("Sensations are brain processes") that is at issue as how they interpret their assertion and what they conclude from it. "Yes", we may reply, "sensations are brain processes and words are letters. But do not interpret this as being a

---

[8] Place, "Consciousness", p. 104.
[9] *Ibid.*, p. 105

mere identity as in the case of the Evening and Morning Star. Sensations (words) are neural firings (letters) but at the same time they are a different sort of thing from what constitutes them. And do not conclude from your identity statement that we could theoretically and/or in practice eliminate talk about sensations (words) and replace it without loss by talk about neural firings (letters).[10] In short, sensations (words) are neural firings (letters) but once understood, this is really *beside the point.* We must not forget that if sensations (words) are constituted by neural firings (letters), then neural firings (letters) can constitute sensations or experiences (words). So it makes little sense to speak here of 'reduction' or 'strict identity' or 'disappearance'. The discovery or claim that sensations are not "ghost stuff" but "brain stuff" is not the discovery or claim that there really are no sensations, any more than the realization that words are letters means that there are really no words."

I have thought it unnecessary to cite all the arguments of Identity theorists designed to show that sensations or states of consciousness are brain processes because I have shown that even if we grant that they are brain processes, we need not conclude that they are 'really' brain processes or that they do not really exist. Also, it seems to me to be evident from such analogies as that of words and letters that A may be constituted by B's and yet be a different sort of phenomenon from B's. And, where we have different sorts of phenomena, it is only to be expected that not every sort of locution will be equally applicable to both of them.

The Identity Theory fails not so much because it is false but because it is irrelevant. This is true of any theory which presupposes that if A is composed of B's then its "true nature" is to be B's [11] or it is B's "and nothing else".[12] It is not true that expressions like "brain process" or "C-fibers firing" adequately characterize the referent of "consciousness" or "pain".[13] It is no more true than the claim that "letters" adequately characterizes the referent of "words" or that the latter adequately characterizes the referent of "statements". As mentioned earlier, to *define* words (sensations) as being letters (brain processes) is to miss the point.

Experiences of pain, unlike the neural processes associated with them, have not just been recently discovered. Only after it was discovered

---

[10] Cf. above, p. 99 and below pp. 128 ff. and 149 ff.
[11] Smart, "Sensations", p. 165.
[12] Place, "Consciousness", p. 103.
[13] *Ibid.*

that sensations are invariably correlated with certain brain processes (assuming that this is the case), was it realized that these neural processes were a hitherto unknown feature of these experiences. This is why it seems peculiar to suggest that sensations are nothing really but processes in the brain; these two phenomena were, and continue to be, independently identifiable. The fact that they are somehow related was established only after it had been determined that they occur simultaneously.

Neurophysiologists could not have begun to make such correlations, they could not have searched for the neurological factors present during certain experiences, unless they could somehow already identify the mental processes (the experiences) independently.

It may be objected that things may be identical, yet have different properties and be independently identifiable. The Evening Star is identical with the Morning Star, they are one and the same thing, yet the former has the property of appearing in the evening and the latter has the property of appearing in the morning. Therefore, it is concluded, experiences too may be identical with neural processes, even though these may not have identical properties and may be independently identifiable.

It seems strange to call "appearing in the morning" a property. It seems to be an action or a relation (to observers) rather than a property. However, even if it is a property, it is not an intrinsic property such as being a particular size or shape. And, if it were an intrinsic property, since only one thing is involved here, viz., Venus (which is called the 'Morning Star' or the 'Evening Star') it seems odd to say that it has different properties from itself. The notion that it has different properties from itself (or the same properties as itself) is meaningless. It conveys no information whatsoever. It is as meaningless as the affirmation that something is like (unlike) itself or identical with (not identical with) itself.[14] Venus could not be independently identifiable from itself. Such a notion is absurd. It happens to appear in the morning and in the evening so for a long time it was thought that there were two stars (or planets) here, not one.

It makes no sense to say that a pain 'appears', since it lacks spatial location in the sense in which physical objects have it. Smart, however, is committed to saying that sensations have all the properties which neural processes have, since, in his view, they *are* neural processes. He seems to accept, what I have been arguing, that if A is identical with B,

---

[14] In his paper "Materialism", Smart says in another connection that "in this essay I am using 'like' in a sense in which a thing is like itself", p. 654.

then it must have all the properties of B. In *Philosophy and Scientific Realism,* he says the following:

On my view sensations do in fact have all sorts of neurophysiological properties. For they are neurophysiological processes....if....the brain process is swift or slow, straight or circular, then the experience *is* swift or slow, straight or circular. For.... the experience is the brain process.[15]

Smart says that sensations have "all sorts of neurophysiological properties" but I presume he would defend the view that they have *all* their properties. He has followed his Identity Theory to its logical conclusion. He realizes that if sensations or experiences *are* neural processes, then their properties must *be* those of neural processes. And Smart is, of course, right. If sensations are nothing but neural processes then they cannot have any different properties from neural processes.

But what possible sense can be made of the suggestion that an experience could be, e.g., circular? This sounds very much like the claim by the early Gestalt psychologists that there is an isomorphism between neural and mental processes such that each different experience has the same structure as its correlated neural processes. Thus if I see a circle, there must be a circular pattern formed by the neural processes underlying my experience. However, this interpretation of what Smart says cannot be the correct one since he claims an identity, not an isomorphism, between the experience and the neural processes.

What could it mean to say that an experience is circular? It could not mean that what I see is a circle since it is the experience itself, not the object experienced, that is supposed to be circular. It may mean that the same experience continually recurs, but this is not what would be meant by a circular brain process. Presumably a brain process is circular if the electrical potentials describe a circle in the neurons. But how would we translate this into a property of an experience? Unfortunately, Smart does not explain how an experience could be circular or straight; he just asserts that it could be.

Since it is meaningless to ascribe such properties to experiences, it must be presumed that they cannot have such properties. This being the case, experiences cannot be neural processes. These considerations apply to all the spatial properties of neural processes. If spatial predicates cannot be applied to something, then it cannot have spatial properties, i.e., it cannot be a spatial thing.

[15] *Op. cit.,* pp. 96-97.

It has recently been argued that there need be no relationship what-soever between experiences, memories or thoughts and brain processes. In *The Brown Book,* Wittgenstein says the following:

> Note....how sure people are that to the ability to add or to multiply or to say a poem by heart, etc., there *must* correspond a peculiar state of the person's brain, although on the other hand they know next to nothing about such psychophysio-logical correspondences.[16]

In the *Zettel* he makes the following remarks:

> No supposition seems to me more natural than that there is no process in the brain correlated with associating or with thinking....Why should this order [of our thoughts] not proceed....out of chaos?....Why don't we just leave explaining alone? [17]

As for "leaving explaining alone", I would prefer to continue looking for neural correlates to thought and sensation than simply to assume that they issue from chaos.

More recently, in an article called "The Correspondence Hypothesis",[18] a follower of Wittgenstein, Bruce Goldberg, echoes his views. Goldberg claims that the correlation hypothesis is nonsense, but he neglects even to mention the experiments and findings of the neurophysiologists, the neurosurgeons or the brain-mappers.

The point is, that it cannot just be a coincidence that whenever a certain part of the brain is probed, a similar (or identical) experience is undergone or that whenever a person is in pain, the characteristic spike pattern appears on the EEG. True, Wittgenstein and Goldberg are mainly concerned with thoughts or behavior and brain processes while the Identity Theory has so far restricted itself to consciousness or sensations and brain processes. However, in the case where the experimenter stimulates the subject's brain and the latter undergoes and/or recounts an experience or incident, it is evident that there may occur simultane-ously experiences, sensations, thoughts and memories. This being the case, a consistent Identity theorist would have to say that all these phenomena are really nothing but brain processes.

For the purposes of this discussion, we have assumed what in any case seems highly probable, viz., that there is an invariable correlation and a concomitant variation between all occurrent mental processes and

---

[16] *Op. cit.,* p. 118.

[17] *Op. cit.,* pp. 106-07.

[18] In the *Philosophical Review,* Vol. 67, 1968.

specific types of neural processes. It is highly implausible that similar experiences should not have similar neural correlates. It is also implausible that it is a mere coincidence that every experience of pain shows up as a spike pattern on an EEG. Where we find correlations like this, we must conclude that there is some relationship between the experience and the brain processes—be it causal, identity or something different from either of these. Both Wittgenstein and Goldberg no doubt were aware that correlations between behavior or experiences and brain processes had been suspected and were being tested. However, they obviously regarded the empirical data as irrelevant and perhaps as untrustworthy. I would suggest that in this case it is philosophical folly to ignore the empirical findings though it is a philosophical duty to exercise circumspection in interpreting them.

Neurophysiologists, as we have seen, tend to interpret this correlation as one between phenomena of wholly different categories, viz., the physical and the mental. There is a dualism, found among laymen and scientists alike, which contrasts physical and mental processes. It is commonly thought that if something is mental it cannot be physical, and if it is physical, then it cannot be mental.

Ryle showed convincingly that behavior could be both physical and mental at the same time, i.e., that something could be an exercise of the body and the mind simultaneously. However, Ryle cautioned against putting the matter in these words because it sounds as though there is a mind which is active and a body which is active. In fact, it is a person who is doing and/or undergoing something and, according to Ryle, this something is mental in so far as it manifests some degree of intelligence or thoughtfulness and, of course, it is physical too.

Being in pain is, despite what Ryle says,[19] a mental process. It is an experience of a certain kind and all experiences are mental states or processes in the sense of 'mental' formerly outlined.[20] But to insist that being in pain is a mental state is not to debar it from being other things as well. This, however, has not been recognized by the Dualists and Identity theorists alike. Both have assumed that if being in pain is a mental state or process then it cannot be physical as well.

The neurologist, W. Russell Brain, asks the following question:

---

[19] Cf. above, pp. 10 ff. Though he usually denies that there is anything mental about pains (Mind, p. 222), Ryle does say at one point that "being in pain is a state of mind" (Ibid., p. 221.).

[20] Cf. above, pp. 11 ff.

Can we imagine how the passage of electrical impulses along certain nerve fibers to an end-station of the brain can result in a sensation of pain? [21]

Brain wonders how physical processes could possibly produce mental ones, e.g., sensations. He conceives the situation *causally*, i.e., assumes that certain neural processes in the brain produce the sensation in the mind. However, as we saw earlier, the causation ends when the brain process correlated with the sensation occurs. Brain's causal dualistic interpretation of these events would be plausible only if there were an independent mental entity or substance (the mind) capable of being affected by brain processes. Indeed, Brain says that he wishes to "preserve the mind's autonomy".[22]

Brain's inability to conceive how such qualitatively different substances or processes could interact is shared by many neurologists.[23] The source of their perplexity lies in their regarding the state of being in pain as purely or exclusively a mental process (a state of mind) which is somehow caused by purely physical processes (a state of one's body). The perplexity is dispelled when it is realized that being in pain involves both experiential and neural features, just to mention two.

Both Cartesian Dualists and Identity theorists err when they identify the state of being in pain with one of its elements, experiential in the case of the former and neural in the case of the latter. In fact, being in pain involves a mixture of heterogeneous components. It is my thesis that a detailed description of what occurs when people are in pain will show that both the Identity Theory and Cartesian Dualism are untenable.[24]

Only a Multi-Aspect Theory of the mind can do justice to our experiences. This theory can account for the fact that there is an intimate relationship between experiences and brain processes without attempting to reduce the former to the latter.

One attraction of the Identity Theory is its purported elimination of inherently private processes from the world. Whereas Archimedes said "Give me a place to stand and I will move the earth", the Identity theorists say "Give me a man's brain processes and I will give you his experience". It is part of my argument, however, that even if we come

---

[21] *Mind, Perception and Science*, p. 66.

[22] *Ibid.*, p. 2.

[23] For example, Penfield; Cf. above, p. 26.

[24] For a contemporary version of Cartesian Dualism, see Jerome Shaffer; "Persons And Their Bodies", in the *Philosophical Review*, 1966.

to the point where we can predict with certainty what a man is experiencing, from our knowledge of his brain processes, the Identity Theory is untenable. A discussion of a dual aspect theory of pain will be followed by an elucidation and defence of a Multi-Aspect Theory of the mind.

We earlier referred briefly to R.J. Hirst's version of the Identity Theory.[25] His view will now be examined in more detail because it points the way toward a more satisfactory theory of the mind than Cartesian Dualism, Rylian Behaviorism or the Identity Theory of Place and Smart. However, his theory does little more than point the way; it needs extensive modification and enlargement if it is to become a plausible alternative to previous theories of the mind.

Hirst outlines his aim thus:

I hope to show that a monistic thesis can be developed which can readily admit mental acts and occurrences, indicate how they are related to brain processes, and account for the privileged access of introspection without making it access to a shadow world of mental substance.[26]

His theory is monistic, and not dualistic, in that it states that a person is not an amalgam of two disparate substances (mental and physical) but "is a unity....and is the self-conscious organism that sees, thinks and acts".[27] He supposes "the mental and....cerebral activities of the person to be two aspects of one and the same activity".[28] He insists that these aspects should not be regarded as "entities or existents in their own right" and he adds the following:

....a double-aspect theory should be monistic in that when one is aware of an aspect of a thing or event one is aware of that thing or event, and so to be aware of two aspects of it is to be aware of the one thing or event from two different points of view, in two different ways, or on two different modes of access.[29]

Thus, for Hirst, the experience of pain which I undergo is "one and the same activity" as the cerebral process in question, but I am aware of it in a different way than is anybody else. Should the objection be made that the experience must be a different event from the brain process because each has some properties which the other lacks, Hirst replies as follows:

---

[25] Cf. above, pp. 57-58.
[26] *The Problems of Perception*, p. 188.
[27] *Ibid.*
[28] *Ibid.*, p. 189.
[29] *Ibid.*

In the problem before us there is clearly a radical difference in modes of access or awareness: the subject's privileged access to his mental acts is by experiencing or introspecting them, but his physical or bodily actions are open to public observation by the senses and appropriate instruments. The importance of this is that a great difference in mode of access will mean a great difference in the characteristics of the aspects revealed, and this may well lead one to regard as two entirely different whole events what are in fact two aspects of one whole event, and so are merely that one event revealed in two radically different ways. The mistake would arise from neglecting the difference in modes of access, and it may well be that this is the origin of the dualist theories: mental and bodily events seem so radically different as to belong to two different orders of being, and so they would be if the modes of access to them were similar; but if the difference is due to the difference in mode of access then they may be two aspects of the one order of events, i.e. be that one order of events differently revealed.[30]

It was argued earlier [31] that my experience could not simply be identical with my brain processes since another person could not *see* what I am *feeling*. Hirst realizes that the differing modes of awareness reveal entirely different features, but he still advocates an 'Identity Hypothesis' according to which "mental and bodily events are *really* identical, only appearing different owing to the different modes of access to them".[32]

Hirst wishes to hold a Dual Aspect—Identity Theory which, I think, is untenable. Either my experience is, or is not, simply the brain process in question. I have attempted to show that even if every experience has a neural factor, or even if every experience is composed of neural processes, it is misleading to say that experiences or sensations are nothing but, are *identical* with, or are *really,* brain processes.

Hirst argues that the "basic whole event or situation" which is constituted by the brain process and the experience is the subject's *being in pain*.[33] He says that "certain mental and bodily events are strictly the same event".[34] From the correlations between sensations of pain and certain brain processes we must, he says, conclude the following:

....the feelings and the appropriate brain activity of a person in pain....are two aspects of the one event in a detailed particularizing sense; hence they are strictly identifiable, i.e., are the one event (his being in pain....) differently observed. A feeling of pain and a pattern of brain activity appear so different that if they were both perceived or both introspected we should have to say that they were

---

[30] *Ibid.,* pp. 190-91.
[31] Cf. above, p. 57.
[32] Hirst, *Perception,* p. 191, emphasis mine.
[33] *Ibid.,* pp. 193-94.
[34] *Ibid.,* p. 193.

two different whole events. But this is not so; their marked difference in appearance can be attributed to the marked difference in mode of access to them. They can therefore be regarded as one and the same event in a person's life, and the difficulties of dualism are thus avoided.[35]

For Hirst, then, my being in pain is the only *event* occurring but I am aware of it as a feeling of pain and another person is aware of it, or could be aware of it, as a brain process. What we are both aware of, however, is one and the same thing, viz., my being in pain. Thus an experience of pain is not an ethereal process occurring in a gaseous medium or substance called the mind but is the inner or subjective aspect of a person's (an organism's) being in pain. And Hirst recognizes the significance of this 'private' or subjective aspect:

....privileged access....has importance in addition to being the source of the inner aspect of these activities: it provides a rough differentia of a mental activity to replace the dualist one that 'mental' means 'in the mind'; one can say that a mental activity is one in which the inner aspect is of primary importance, i.e., one where privileged access reveals *more than* external observation and provides the features which distinguish the activity from others.[36]

While I regard Hirst's theory as an improvement on other Identity Theories, I think that it embodies at least one of their misconceptions. This concerns the claim that we have a strict identity here and not just two aspects of one event.

Hirst says that what I am aware of when I am in pain *is* what you see, i.e., my cerebral process. He says that one's privileged access reveals *more* than "external" observation. I think it would be less misleading to say that what I experience could be seen by nobody, since it is not that *sort* of thing, and that what is "revealed" to me during my experience is not something *more* than the cerebral processes (since these are not revealed at all to one in pain) but is one of their features or aspects.

Hirst's contention that my experience and my cerebral processes are identical (one and the same activity) is not unlike the claim that words are really letters or that tables are really elementary particles. If experiences or sensations (tables:words) and cerebral processes (elementary particles:letters) were one and the same, or identical, then one's being aware of the former would mean neither more nor less than one's being aware of the latter.

[35] *Ibid.,* p. 194.
[36] *Ibid.,* p. 196, emphasis mine.

Perhaps it will be argued that one can be aware of X without being aware of all the aspects of X or of the constituents of X. For example, somebody may see one side of a mountain and, though not observing or being aware of its other side, would still be said to *see the mountain*. Thus it may be argued that sensations could be brain processes or be aspects of the same thing as them, even though the one having the sensation is not aware of his brain processes as such. The fact that one in pain is not aware of his brain processes as such, it is concluded, does not exclude the possibility that his pain *is* a neural process or an aspect of one and the same thing of which the neural process is an aspect. Thus sensations may be one and the same with one's brain processes even though one is unaware of the latter. In other words, one can be aware of one's pain without being aware of certain aspects of it and/or without being aware of its components. Sensations may be brain processes even though those undergoing the former are unaware of the latter as such.

I agree that we may be aware of X without being aware of some of its aspects or of its components. One can see a table without seeing the bottom part of the legs and without seeing the molecules, atoms or protons and electrons which constitute it. All this is obvious. But what if it is said that tables or sensations are 'identical with,' 'one and the same as', or 'really' the ultimate particles which constitute them? This is a *different* statement from the true one that we can be aware of something without being aware of all its aspects or its components. It amounts to the claim that *what we are aware* of is nothing *really* but these ultimate constituents. And this means that we are actually aware of these ultimate constituents.

Nobody would claim to be aware of his own brain processes when in pain or to be aware of electrons and protons when seeing a table —even if these phenomena are constituted by the elements in question. Nor would anyone claim that the other side of the mountain is what the mountain *really is*. The notions of aspects and of constituents are different but they have this much in common, viz., one can be aware of X without being aware of all its aspects or of its smallest components. Also, it is as mistaken to identify something with one of its aspects as it is simply to identify it with its components.

I admit that brain processes can constitute our sensations even though we may not be aware of our brain processes as such. But I reject the thesis that sensations are nothing but brain processes because this implies that the neural processes do not constitute sensations. My experience of pain may be composed of brain processes or may be one aspect of a phenomenon of which they are another aspect but this does not mean that my experience is not real or that I am not really aware of my experience but of brain processes. And Hirst *does* say that "certain mental and bodily events are *strictly the same event*." [37]

It would be misleading to say that what another sees (my brain processes) and what I feel (my pain) are simply one and the same thing. When one sees the Morning Star or the Evening Star one is seeing Venus in the morning or in the evening. When one sees brain processes which constitute an experience of pain, one is not seeing an *experience* of pain. And there is a sense in which one is *not* aware of cerebral processes (elementary particles:letters) when one is aware of pain (tables:words), unless, of course, one is somehow observing one's cerebral processes as one experiences their correlated pain.

Wittgenstein, as we saw, envisages just such a subject-experimenter in the *Blue Book*. He is experiencing pain [38] and is somehow seeing his own brain processes. Wittgenstein argues that he "is observing a correlation of two phenomena". He claims, I think rightly, that it is unhelpful to "say that he is observing one thing both from the inside and the outside; for this does not remove the difficulty".[39]

This is why it is unhelpful. If someone were looking at a table, it would be misleading to tell him that what he is really seeing is, or what he is seeing is really, elementary particles. We have already argued at some length against the view that tables or words are *really* elementary particles or letters. Granted that these arguments are cogent, then it would be, perhaps, trivially true to say that one seeing a table is seeing what is in fact composed of elementary particles, but it is misleading if this is supposed to mean that he is aware of the elementary particles as such.

When someone observes a table, is he observing the elementary particles out of which it is constituted? To say "No" might be interpreted as a denial that the table is composed of elementary particles. To say "Yes" might be interpreted as the claim that people commonly observe elementary particles, which is patently false. The truth of the matter is that he does not observe, is not aware of, any elementary particles but that what he is observing (the table) is composed of innumerable such particles. With regard to experiences and cerebral processes, one in pain

---

[37] *Ibid.*, p. 193, emphasis mine.

[38] Wittgenstein uses thoughts instead of pains in his example but the principle is the same.

[39] *Op. cit.*, p. 8.

is not aware of one and the same activity (the brain processes) as an observer. And if experiences or sensations were *composed* of the firing of certain neurons in the body, this would still not mean that the one in pain and an observer were aware of one and the same phenomenon.

Hirst maintains that the subjective aspect of pain (the experience) is "really identical" [40] with the neural component of pain. Both constitute, according to him, what he calls the whole event, viz., being in pain. He claims that the one in pain and an observer both have access to one and the same thing, viz., the pain. Being in pain involves both experiential and neural components and access to either of these components gives us knowledge of the pain which the person is undergoing.

In the discussion of someone observing the brain processes of a subject experiencing pain,[41] and elsewhere,[42] it has been argued, in effect, that there is a sense in which they are aware of the same thing but that in a more important sense they are aware of different things, viz., the neural processes and the experience. Hirst says that they are aware of different features, namely, an experience and brain processes. He prefers to speak of the neural-experiential phenomenon as an event in itself and to speak of the subject and observer as being aware of the two different features or components of the pain. He thus combines an Identity Theory with a Dual Aspect Theory, regarding the pain as a psychophysical event with an inner aspect (experiential) and an outer aspect (neural). By 'aspect' Hirst means 'view' [43] and so we have one event (the pain) being viewed from the inside by the subject and from the outside by an observer. Each is aware of one component of the total phenomenon.

This view is reminiscent of Spinoza's wherein the physical and the mental are actually one and the same phenomenon being viewed now in one way, now in another. Hirst differentiates his theory from traditional double-aspect theories which, he says, postulate a third substance of which the mental and the physical are but aspects.[44]

If we deem pain to consist of the total neural-experiential phenomenon, then Hirst is right in saying that the subject's access and an observer's access are both access to one and the same thing, viz., the pain. But if we were to adhere to what seems to be the prevalent notion of pain

---

[40] Hirst, *Perception,* p. 191.
[41] Cf. above, pp. 39 ff., 51 ff., and 55 ff.
[42] Cf. below, pp. 118 ff.
[43] Hirst, *Perception,* pp. 189-90.
[44] *Ibid.,* p. 189.

as seen in the common usage of the word 'pain', we would say that an experience is a different sort of phenomenon entirely from brain processes. In this case, the observer's access is just to brain processes and the subject's access is just to his pain (experience), even if the experience should happen to be *constituted* by the brain processes.

It seems to me that Hirst is right in regarding pain as a psychophysical event having both experiential and neural components, given that pain does have a neural factor. At the same time, it must be made perfectly clear that what is normally understood by the word 'pain' is the experiential component alone. This is why the Identity Theory of Place and Smart, *et al,* is incredible. It is one thing to identify the experience of pain with the neural processes in the sense that they are two components of one and the same psychophysical event, viz., the pain one is in. It is another thing entirely to identify the experience with neural processes in the sense that the experience simply *is* the neural process. Smart specifically repudiates any sort of Dual Aspect view.[45] In effect, he identifies the experiential *component* of pain with its neural component whereas Hirst rightly differentiates between these—though he sometimes says that they are "strictly identifiable".[46]

What Hirst does, in effect, is to incorporate what *constitutes* pain into the *concept* of pain or the meaning of 'pain' (or 'being in pain') by regarding the state of pain as the total psychophysical event rather than as just an experience.[47] In fact, this might be less misleading, e.g., to the neurophysiologists discussed earlier, than to continue to regard pain simply as an experience which is mysteriously related to certain brain processes. When using 'pain' as commonly understood, we refer to the experience (experiential feature) alone. Given that it is determined that pains are constituted by brain processes, we *might* come to use 'pain' to refer to both of the components invariably occurring in the person in pain. (And being in pain may characteristically involve other features as well.)

We may call the experience of pain a different sort of thing from neural processes, even though it is constituted by them, or call it a different sort of component from the neural one. If we admit that the experience of pain may be constituted by neural firings without being reducible to them, it does not matter so much how we phrase this duality as it does that we acknowledge it.

[45] Smart, *Scientific Realism,* p. 94.

[46] Hirst, *Perception,* p. 194.

[47] Cf. above, p. 46.

Other Identity theorists (and Ryle) discussed in the course of this book tend to minimize, if not explicitly deny, the 'privileged access' which persons have to their own sensations. Hirst's view represents a considerable conceptual advance over the Identity Theory of Place and Smart because it rightly emphasizes the asymmetry between self-knowledge and the knowledge of others regarding pains. Also, Hirst recognizes that the experiential aspect [48] is the *distinguishing* one of these mental processes and that it is of primary importance. My 'privileged access' is access to something to which nobody else could have such access, viz., my experience. Another may have access to the neural component of my pain but never to its experiential aspect. And my 'access' is not to the cerebral aspect but to the experiential aspect. (I use the term 'aspect' here to mean 'feature'—not 'view'.)

Hirst's view is highly suggestive and is an improvement on the Identity Theory. But he finds only two aspects (views) of pain whereas I shall outline several aspects (features) of pain.

Not only is it false, then, that an experience and a brain process —whether seen as events in their own right or as aspects of a psychophysical event—are "really identical" but they are different *sorts* of phenomena or features. The same is true of elementary particles and tables. These too are different sorts of phenomena. Tables may be green or brown but no sense attaches to the notion that an electron is green or brown. And yet, a table is composed of elementary particles. As we saw earlier, there is nothing paradoxical in the view that S may be composed of B and yet neither simply be identical with B nor have all the sorts of properties which B has. Even if we grant that experiences *are* neural processes, in the sense that they are composed of them, it would not follow that they are identical with, or are nothing but, these neural processes.

It may be objected that there is a sense of identity according to which the neural and experiential components *are* parts of the identical thing, i.e., the pain which the person is undergoing. The following examples might be used to illustrate the point.

If we speak of a person's arms, legs and head we are talking about a person, one thing. When we speak of the leaves, branches and sap of a tree we are again speaking of one thing, the identical thing, viz., the tree. Thus we can speak of the parts or properties of something and still be speaking about the identical thing.

---

[48] Or "inner aspect" as he calls it on p. 196 of *Perception*.

I see nothing in this which contravenes anything I have said. I agree that it might be less misleading to consider the neural and experiential factors together to constitute pain. In this case, pain would be a whole event having two components, neural and experiential. And these two factors would be parts of the *identical* event, viz., the pain.

But an Identity theorist (not necessarily Hirst) identifies these two factors. He does not make the uncontroversial claim that the neural and experiential features are parts of one and the same (or the identical) whole. He makes the provocative claim that the experiential factor or process is nothing really but the neural factor or process. And this is like saying that the head of a person is nothing really but the arms or that the leaves of a tree are nothing really but the branches! The Identity Theory is nothing if it is not *reductionist*. Its exponents characteristically appeal to Occam's razor and there would be no call for a razor if there were nothing they wanted to eliminate.

When Identity theorists speak of 'identity' they mean the identity of the sensation with the neural processes. This is the view which I am rejecting. When Hirst modifies the Identity Theory so that the experiential and the neural factors are seen as parts of the *identical* phenomenon —the pain—I agree that it might be more helpful to regard pain as a total psychophysical event having subjective and objective aspects.

So long as the duality between what I feel when in pain and what another sees is maintained—whether these be called processes, or factors of one and the same (or the 'identical') process—I see no reason to object to Hirst's formulation. I do think he might have called his theory something different from an 'Identity Hypothesis' because this suggests that he is a reductionist in the same way that Identity theorists are reductionists.

Whether being in pain is regarded as an experiential state in its own right or as part of a total psychophysical state, it cannot be identified with or reduced to the neural processes or factors. Place, Smart, Armstrong and Feyerabend wish to identify the experience with the brain processes. Hirst wants to call the experience a feature of the identical process of which the neural processes are a feature. These seem to me to be completely different views. The former suggests that all that is really transpiring when someone is in pain are certain neural processes. The latter view admits that being in pain involves an experience or experiential feature. I reject the first view as being very misleading and perhaps incoherent. I shall now develop the second view into a Multi-Aspect Theory of the mind (and of sensations).

CHAPTER SIX

A MULTI-ASPECT THEORY OF THE MIND

Using the example of someone suffering from a severe toothache, we shall now propose a Multi-Aspect Theory of the Mind which is more comprehensive than any so far discussed and which avoids the difficulties of Cartesian Dualism, Rylian Behaviorism and any sort of Identity Theory. Some components of this theory have already been suggested but here we shall describe it in detail and defend it.

It should be noted at the outset that almost all the proposed definitions of 'mind' advanced by philosophers have been inadequate or utterly misleading. Paraphrases of some such proposals are the following:

> The mind is a thinking thing — Descartes.
> The mind is one's ability and proneness to do certain sorts of things — Ryle.
> The mind is the brain — Armstrong.

A more accurate definition is that of R.J. Hirst. He suggests the following formulation:

> ....when we speak of a man's mind we are not speaking of some entity lodged in his body or of a mental substance, but we are referring to the pattern of his mental abilities and dispositions, and so to the ways he can and will react, not necessarily overtly, in various circumstances.[1]

On the previous page, he says that 'mental' is "being used in its ordinary contrast to 'physical' and referring to thoughts, feelings, perceptions, volitions or intellectual powers".

Thus Hirst realizes that 'the mind' refers to both experiences and ways of acting and also that being in pain involves both neural and experiential aspects. In the course of presenting a Multi-Aspect Theory, it will be

---

[1] Hirst, *Perception*, p. 187.

argued that being in pain may involve as many as five aspects, not two as suggested in Hirst's Dual Aspect-Identity view.[2]

Suppose, then, that a person (X) is suffering from a severe toothache. We can see his decayed tooth and, let us say, we can observe all the other physiological, including neural, factors involved in his experience. X is feeling, we may suppose, a considerable degree of pain, an amount which is consonant with the extent of the decay of his tooth.

It might be said by a Cartesian Dualist that X's feeling of pain is a mental process which is (somehow) being caused by the decayed tooth. The decayed tooth, on this view, is causing electro-chemical impulses to be sent to the brain via the rest of the nervous system and these in turn cause certain brain processes to occur. The Dualist concludes that it is these brain processes that are (somehow) causing the (mental) sensation of pain.

This is the picture of pain, and other such processes, which many neurophysiologists seem to have. In a recent seminar, many such scientists express their utter ignorance as to how a process in the body could produce a process in the mind.[3]

A less perplexing interpretation of the facts would be that being in pain is not merely a mental process which is mysteriously produced by a brain process but is a psychoneural process involving at least an experiential and a neural factor. It is not the case that a physical process in the brain causes a mental process in the mind but that a brain process and an experience occur simultaneously. Whether the experience of pain is produced by hypnotic suggestion or a decayed tooth, it is a psychophysical event involving at least two factors, the experiential (psycho) and the neural (physical).

Man is not a mind temporarily ensconced in a body. Nor is he nothing but a very complicated physico-chemical mechanism. He is a psychophysical organism, many of whose states and activities are classified as 'mental' for one reason or another. What makes a chess move mental [4]

---

[2] In the sense that there are inner (subjective) and outer (objective) features of pain, we might say that there are two aspects (views) of pain but aside from this epistemological dualism, being in pain can involve many different kinds of features.

[3] *Brain and Conscious Experience,* ed. by John C. Eccles, Springer-Verlag, New York, 1965.

[4] A chess move might be classified as 'mental' because it embodies learned rules and strategy. Since the ability to learn and plan is a criterion of intelligence, which in turn is a criterion of mentality or mind, we could say that a chess move is 'mental'. Intelligent behavior bespeaks some degree of mentality.

need not be what makes an experience of pain mental. What are needed are detailed descriptions of the different types of states and/or activities which would be classified as 'mental' and, of course, much more investigation into the workings of the brain.

It has been wisely said that "The human body is the best picture of the human soul" [5] but it would be folly for philosophers to ignore the brain research that is being conducted. Our understanding of such phenomena as memory and perception should be greatly enhanced when their neural correlates are discovered and studied. This does not mean that philosophical analysis of the concepts of memory, pain or perception are fruitless or that phenomenological descriptions of pains and other phenomena are pointless. It just means that our understanding of some of the physical factors involved in, e.g., pain, may help us to understand better some of its other factors.

With regard to X and his toothache, Hirst would interpret his being in pain as a process involving an inner, subjective view of the pain (X's view of it) and an outer, cerebral view of the pain (everyone else's view of it). An adequate analysis of the subject's *being in pain,* however, would have to take into account the following aspects or features: experiential, neural, bodily, behavioral and verbal. We shall now take a look at each of these in turn.

1. The experiential aspect: This is the felt quality or the sensation of pain—the "raw feel" as it has sometimes been called. It is what one experiences or feels when one is in pain. This is the feature of pain which leads us to call pain a 'mental' process. It is because this feeling or sensation could not exist unfelt or unsensed that we call it 'mental'. That is, it could not be truly said to *exist* unless it were sensed; "I am in pain but I am not aware of the pain" is self-contradictory. It could not exist unless it existed as an aspect of consciousness.

This is the aspect of being in pain which is referred to in the everyday use of the word 'pain', e.g., "I am in severe pain".[6] It is what is being complained about when one shrieks in agony. It is the feature of being in pain which Identity theorists have, not unnaturally, been taken by some to be denying when they claim that sensations are identical with, are nothing over and above, or are really, brain processes.

---

[5] Ludwig Wittgenstein, *Philosophical Investigations,* p. 178.

[6] In everyday life, of course, we do not normally announce to others in a declaratory tone of voice that we are in severe pain.

It would not be misleading to say that a satisfactory answer to the question "What is pain?"—as it arises in a philosophical discussion—is that it is the sensation or experience (experiential aspect) under discussion. The feeling or sensation of pain, i.e., what the individual feels, is the *sine qua non* of being in pain. This is why it is self-contradictory to say that someone is in (feels) pain but he does not know it or is unaware of it (does not feel it).

D.M. Armstrong disagrees: "Now we can have a sensation of pain and be perfectly unaware of having it. So there can be a feeling of pain that we are unware of feeling: unconscious pain"[7]. If one could be in pain and not know it or be aware of it, it would not make sense to ask the individual how much pain he was in, what sort of pain it was or even where the pain was. Since these questions always make sense when someone has a pain or is in pain, it would seem to be impossible to have a pain of which one is totally unaware. We would not *call* something 'pain' which was unconscious. How could one claim to be in pain in the first place if one were totally unaware of it? What would lead the one 'in pain', or others, to think that he *was* in pain unless he displayed it in some way and/or told them about it—in which case he would not be unaware of his pain? The following statement seems to me to be incoherent: "It hurts (or "I'm in pain....) but I have no idea how much it hurts or where it hurts (or even *that* it hurts)". If we cannot sensibly say such things, it is because the notion of an unconscious pain is vacuous. It would seem to be impossible to distinguish between an excruciating unconscious pain and a slightly distressing unconscious pain but it is *always* possible to distinguish between these two degrees of pain as it is normally understood. (I do not claim that there could be no unconscious mental states or activities, but I claim that the idea of an unconscious *pain* is vacuous.)

It might be argued that we can distinguish between being in pain or having a pain and *knowing* that one is in or has a pain. For example, animals other than man might be said to be in pain or have a pain but not know it. They do not know it, so the argument goes, because they have no concepts, including that of pain.

Do animals in pain *know* that they are in pain? If a straightforward "Yes" or "No" is demanded here, then this question is unanswerable. If the questioner wants to know whether animals are to some degree aware of their pain, then we could answer in the affirmative. But if he

---

[7] *A Materialist Theory of the Mind*, p. 312.

wants to know whether animals know what pain is or whether they have a word for pain, a concept of pain or an understanding of pain, then we could answer in the negative. (The dictionary reports both of these usages of 'know', among others.)

Suppose that someone requests to be allowed to breed dogs for the purpose of torturing them in order to discover certain things about their central nervous system and their reactions to extreme pain. Or, he might want to torture dogs simply because he enjoys doing this. In either case, he argues that there is nothing wrong with torturing dogs because, since they have no concepts and therefore no knowledge, they cannot be said to know that they are in pain. If they do not *know* that they are in pain, he concludes, what could possibly be wrong with torturing them? ·

One might as well argue that we ought to torture babies or idiots on the grounds that they have no concepts and so no knowledge of anything, e.g., whether they are in pain or not. And if we make it a condition of knowing something that one have concepts and, presumably, a language, then it seems likely that babies, idiots and dogs do *not* know that they are in pain. Therefore, aside from the bodily damage done to potentially useful beings and aside from the troublesome noises they make when tortured, it might be argued that there is no objective reason to refrain from torturing them. (What they don't *know* won't *hurt* them.)

In reply to such an argument, we must make the distinction between knowing something in the sense of understanding it, classifying it, having a concept of it and a word for it and, on the other hand, knowing something in the sense that one is to some degree aware of it. In the sense of 'know' that counts here (and throughout this discussion), i.e., the second sense, babies, idiots and dogs know that they are in pain.[8] Of course none of them may 'know' anything in the propositional sense that they believe things, with good reasons or adequate grounds, which are true. But to say that they might be in pain without knowing it implies that they might be in pain and be totally unaware of it. Were this the case, it would be puzzling why we can condition or teach animals to fear certain stimuli which have been associated with painful experiences. Why should a dog shun such stimuli if he were totally unaware of pain during the earlier experiments?

One might argue that the dog was not aware *that* he was in pain, even though he was *aware of* the pain. I agree that animals besides man cannot verbalize their pains as we do. If they could, there would likely

---

[8] Cf. below, pp. 145 ff.

be fewer experiments conducted on animals. And if they cannot *tell* anyone that they are in pain, this could be interpreted as implying that they might not know that they are in pain in the propositional sense of 'know'. But this sense of 'know' is not relevant here. To admit that animals are in some sense and to some degree *conscious* of their pain is to admit the possibility that they could not be totally unaware of their pain *whenever* they are in pain. I argue that for any sentient organism to be in pain or to have a pain means that it must in some sense and to some degree be conscious of the pain. (Could a dog, baby or idiot *mind* having pains unless it (he) were somehow conscious of it?)

It would make no sense to deplore the infliction of pain on animals, aside from the two reasons given above, if they in *no* sense knew they were in pain. It seems to be impossible to drive a wedge between being in pain and being in some sense conscious of the pain. This is indicated by the absurdity of this statement: "I am in pain but don't worry— it's an unconscious pain".

If sensations were nothing really but brain processes, it would not be obviously absurd to suggest that one could be in pain (undergo certain brain processes) yet be unaware of it (them), i.e., not feel any pain or be aware of undergoing any pain (brain processes). There are numerous processes in the brain which, so far as we can tell, are not features of any states which involve experiences, i.e., they do not play a part in any state of which the individual is conscious. We have a way of distinguishing between those brain processes which are a factor in states of pain from those which are not, viz., the former are the ones which invariably produce a spike pattern whenever we are in pain. To say that sensations are nothing but brain processes gives the mistaken impression that the brain processes in question do not differ significantly from any other brain processes. However, the crucial fact about the neural processes involved in pain is that they play a part in an *experience*. That is, we have learned that these brain processes are involved in the experience of pain because we can identify or pick out states of pain independently of identifying the relevant brain processes.

Being in pain, then, is undergoing a certain sort of experience in which there is invariably a neural factor involved. There may well be a neural factor involved in *all* our experiences but we will discover these only by investigating what occurs in the body, including the brain, when individuals undergo these different experiences.

Just as we are discovering the neural correlates of different sorts of behavior, we shall, perhaps, discover the neural correlates of different sorts of experiences. And it would be as misleading to identify behavior with the physiological phenomena as it is to identify experiences with them. Just as the same physical phenomena may constitute different behavior in different contexts, so the same physical phenomena may constitute, or play a role in, different experiences in different contexts.[9]

X's experience, then, is what we might call the 'mental' aspect of his being in pain. But to call this feature 'mental' does not mean that it is a state undergone by a mental entity or substance called 'the mind'. A person, not his mind, undergoes experiences. What makes us call the experiential aspect of being in pain 'mental' is its peculiar epistemological status, viz., the one in pain needs no criteria or evidence in order to know that he is in pain. His having the pain, his being aware of the pain, constitutes his knowing that he is in pain. As we noted before, his experience or feeling of pain could not *exist* unexperienced or unfelt. ('Pain' refers to a sort of experience.)

The aspect of being in pain under discussion, the experiential or peculiarly 'mental' aspect, is the only aspect of the pain itself known to the vast majority of people. How pains feel, what it feels like to be in pain, is what is known and what counts to people.

We have granted that being in pain involves a neural factor and that, since the neural processes are invariably correlated with pains, the latter could not exist without them.[10] Eliminate the neural factor and the experiential aspect is also eliminated. On the other hand, if a person's attention is distracted from his pain, the pain (experiential aspect) may cease and along with it the neural processes. Both aspects are integral parts of the whole process of being in pain which never exists without at least these two factors being present. They are like the sides of a coin which could not exist without having both sides.[11]

---

[9] See above, pp. 35-36, for a report of findings which raise the possibility of *identical* physiological phenomena being associated with *different* experiences, when these phenomena occur under different circumstances. Whether the neural processes *are* identical during the different experiences has, so far as I know, still to be discovered.

[10] By the same token, if all pains have a neural correlate, the latter could not exist without the pains existing. Presumably, when a hypnotist convinces a subject that he is in pain, the moment he feels the pain there will occur simultaneously appropriate neural processes. This is why it is futile to ask whether the neural processes *cause* pains or vice versa—they occur simultaneously whether they are induced by a brain probe or by suggestion.

[11] The difference is that whereas it is inconceivable that there could be a coin which was normal in all respects except that it had only one side, it is not

I am making an empirical claim, not a logical one, in asserting that if one's attention is distracted, the neural processes associated with the pain will cease as and when the pain ceases. It is logically possible that we might have found no neural correlates of pain whatsoever. Since we are discovering the neural correlates of various sorts of behavior and experiences, it seems reasonable to assume that the appearance or cessation of instances of the latter will be accompanied by a like activity of the former.

It may be asked how we know that the experiential and neural aspects *are* integral parts of pain—not the concept of pain [12] but pain itself —especially if we differentiate between these two factors. The answer is that we know that each of them is part of being in pain—not part of what it now means to be in pain, i.e., not part of the current concept of pain—because in all the neurological studies of pain to date, a neural factor has been present. Neural processes do not figure in our present concept of pain; so far as this concept is concerned, we might have discovered ghostly correlates or no correlates at all for our pains. If a friend complains of backache, we know what he means—whether the ache is composed of neural processes, ghostly processes, demons or of nothing at all.[13]

It is now becoming common knowledge that pains and other experiences have neural correlates as opposed, say, to demonic or ghostly correlates. If one considers pain to be caused or constituted by demons, then demons rather than neural processes *may* figure in one's concept of pain. But whatever we replace demons with, people will continue to feel pain; whatever causes or constitutes pain, it will continue to be a distressing experience. We are no more justified in simply identifying pains with neural processes than we would be in simply identifying them with demons. To identify the causes or constituents of something is one thing; simply to identify that thing with either of these is quite another.

X has, then, a severe toothache. He is undergoing an experience of pain in his tooth. Experiences are called 'mental', as has been argued, because one cannot have an experience without being *aware* of it. They

---

inconceivable (though it might be false) than an experience might occur which had no neural correlates. We can conceive of someone undergoing an experience of pain without the neural processes occurring. Of course this will never happen if, as seems to be the case, experiences are invariably correlated with processes in the brain.

[12] Cf. above, pp. 116 ff.

[13] A similar point is made by M.J. Budd in "Materialism and Immaterialism", in *PAS,* March 23, 1970.

are also called 'mental' because the individual needs no criterion or evidence in order to determine that he is in pain—he just experiences it. There is yet another feature common to all experiences which does not seem to be common to all physical states or processes. This is the feature which is commonly called 'intentionality'.

One could not have an experience unless it were an experience *of* something. Experiences are referential—they refer to or 'intend' an object. In regard to pains, one may have an experience of pain in one's tooth, one's back or one's stomach. Or, one's whole body may ache. But it would not make sense to say that X was undergoing an experience that was an experience of nothing. To say that X is experiencing nothing is to say that X is not having an experience.

The same applies to belief. One could not have a belief which was about nothing. Nor could one have an idea or thought which was not an idea or thought of something. States of awareness or consciousness must also be *of* something. One could not be said to be conscious unless there were something of which one were conscious, even if one were conscious of nothing in a room but of one's presence in a room or of one's surroundings.[14]

Whether the object of consciousness be something external (a book), inside oneself (a pill sliding down one's gullet) or a pain or an itch, there must be something of which one is conscious.

Also, the object of consciousness may be events which occurred in the past, whether they be activities once engaged in, people once known or pains once felt. Or, one may be worrying about one's future, fearing what tomorrow will bring or hoping that the weather will improve.

In all these cases, it is probable that each experience or thought will be invariably correlated with specific processes in the brain. However, we have now found another feature of mental states, processes, acts or dispositions which is not shared by brain processes. In what sense could a process in the brain be said to be 'intentional' in the sense under consideration? In what sense could worrying about the weather ruining one's crops be *really* a process in the brain?

If such locutions as "I'm worried that the weather may ruin my crops" or "I remember her standing there" are not translatable without loss into such utterances as "My W-fibers are firing" or "My M-fibers are firing again" then there must be a good *reason* for this being so.

---

[14] I have in mind here the case of a person who is completely paralyzed but who is conscious of his surroundings.

It is highly significant that Place and Smart[15] are careful to insist that talk about mental states may not be replaceable by talk about brain states. If it is true, as it seems to be, that we *cannot* eliminate (without loss) talk about mental processes, states, acts or dispositions, it can only be because these are not simply processes in the brain. In other words, if sensations, for example, were nothing *really* but brain processes, then there could be nothing *really* lost by replacing talk about sensations with talk about these brain processes.

In an article called "The Neural-Identity Theory and the Person", Errol E. Harris expresses the point lucidly.

What sort of identity then is strict identity such as it is said to obtain between the mental and neural activity. We are told that it is an identity of the referent of two different languages, the physiological and the phenomenological. But can this be sustained? I do not think so.

If two different languages have the same referents, there must be some similarity in their logics. They must be mutually convertible in some way, as mathematically isomorphic calculi are mutually convertible. This is not the case with phenomenological and physiological languages. 'I remember, I remember the house where I was born' is not convertible into 'There are such and such neural discharges occurring in my brain'; for to remember something is to distinguish and identify a past event, whereas the passage of ions along nerve fibres makes no *reference* to past events whatsoever. To say merely that the sentences do not mean the same because each language has its own peculiar logic may be true, but if so it is to that extent at least *prima facie* evidence that they *refer* to quite different subjects and not to identical events.

If the terms do not mean the same, if what they describe have no properties in common, if their logics are diverse and the languages to which they belong are not mutually translatable, on what grounds do we identify their referents? What evidence of identity is there? In these circumstances, can 'strict identity' have any significant meaning? [16]

Sensation-talk cannot be translated without real loss into brain process-talk nor could both of them be translated into some third type of expression. Whereas consciousness or experience could not exist independently of something of which it is consciousness or experience, it does not seem even to make sense to say that a process in one's brain could 'have an object' or be 'intentional'.

Experiences, memories and thoughts often have references and mark distinctions *in the world* outside of one's brain or body. It is difficult to

---

[15] Place, "Consciousness", pp. 102-03; Smart, "Sensations", p. 162.

[16] In the *International Philosophical Quarterly*, Vol. 6, 1966, pp. 521-22, emphasis mine.

conceive of how a process in one's brain could do these things. Such references and discriminations do not seem to be explicable and analyzable solely in terms of brain processes. How, for example, could the thought of my wife or the experience of seeing her, be understood solely in terms of brain processes? Surely some reference would have to be made to *her* even if the thought or experience were nothing really but a brain process.

It is even more difficult to imagine how a profound experience could be understood solely in terms of brain processes. For example, the feeling of extreme closeness with another person, whether established through long association and mutual affection or through the taking of LSD together could not be explained solely in terms of the firing of certain neurons. And if experiences cannot be understood solely in terms of the firing of certain neurons, it is difficult to see how the experience could be nothing really but the firing of those neurons.

We conclude that this aspect of X's being in pain, the experiential aspect (the experience), is the distinctively 'mental' feature. But there is more to being in pain than just experiencing pain.

2. The neural aspect: These are the brain processes which we have granted to be invariably correlated and concomitantly variable with specific experiences. As neurophysiologists will readily admit, extremely little is known about how the brain operates and very little investigation has been conducted into the phenomenon of pain.[17] However, it was thought that criticisms of, and an alternative to, the Identity Theory, would be more readily accepted if as much as possible were granted to it.

The firing of the neurons in question, then, has been conceded to be necessary and sufficient for the occurrence of the experience. If the appropriate neurons were not firing in X's brain, and the rest of his body,[18] then he would not be having this experience.[19] This aspect too is a *sine qua non* of being in pain, but this is something discovered (we are conceding) only recently. Nobody would speak of 'pain' where there

[17] It should be obvious why it would be preferable to avoid conducting experiments on pain unless they are absolutely necessary.

[18] There are neurons throughout the nervous system so that it is misleading to say even that experiences are correlated with *brain* processes. Except in cases of direct brain stimulation, then, experiences will be correlated with *neural* processes.

[19] As mentioned above on p. 126, if X were not having the experience, these neurons would not be firing.

was no experiential aspect present, even if the brain process covariant with the sensation somehow occurred by itself.

The day may come when we would refuse to acknowledge the existence of pain unless the appropriate neural processes are occurring.[20] Just as we may invent the perfect lie detector—if anything perfect or infallible *could* be invented—so we may invent the perfect pain detector. Such an instrument would not be useful in detecting pains in the borderline cases, however. Since pain functions as an indication of bodily damage, if we are not paralyzed or otherwise incapacitated, the degree of bodily damage will often (though not always) [21] be registered by an appropriate degree of felt pain. If a person were undecided as to whether his experience should be called a 'pain' or simply a 'discomfort', detecting the neural processes themselves will not be of any help. If experiences are nothing but brain processes, then when the *former* are ambivalent, the *latter* must also be ambivalent.[22] If experiences are invariably correlated and concomitantly variant with certain neural processes, then it seems evident that borderline cases of pain and discomfort will have corresponding ambivalent neural states. Thus, while the 'perfect' pain detector would seem to be no better than the *human* detector (the one in pain or dis-comfort), its existence would at least confirm the Correlation Hypothesis,[23] if not the Identity Hypothesis.

In addition to X's experience of pain, the experiential aspect, there will be occurring neural processes. These processes constitute the second feature of being in pain. These are the two factors which together are necessary and sufficient for the occurrence of pain. In cases of direct brain stimulation, these two factors may exist alone. Being in pain, in such cases, may involve nothing more than an experiential and a neural factor. There are, however, three other factors which are usually part of being in pain but which are not necessary, either singly or in concert, for the occurrence of pain. One or the other of them does play a role, nevertheless, in our concept of pain just because they usually do occur when pain does.

---

[20] Cf. above, pp. 90 ff.

[21] A cerebral thrombosis is likely to be lethal though the pain which it causes may be, I believe, minimal.

[22] Even when adopting the language of the Identity theorist, some sort of dualism makes itself felt. This suggests that the Identity Theory is not a coherent, consistent theory.

[23] This is the hypothesis that all experiences, and perhaps thoughts also, have neural correlates.

3. Bodily[24] aspects of pain: An example of this aspect would be X's decayed tooth. Of course a decayed tooth is not, properly speaking, an aspect of *mind*. It is, however, very often a feature of toothaches—states which are characteristically *mental* in the senses outlined above. Thus the bodily aspects of *being in pain* are part of our concept of pain. When someone is in pain we naturally assume that there is something wrong —some physiological irregularity due to damage or disease which is the source of the pain. We naturally assume it because there characteristically *is* something amiss physiologically when one is in pain.

In the case of X and his toothache, it is most unlikely that we would know about the decay unless we first knew about X's pain, except if the decay were visible whenever X opened his mouth. And this will normally be the case whenever the physiological irregularity is inside the person's skin or in a bodily orifice. In cases where the damage is external, there may be a rupture of the skin or some other visible sign. When the source of the trouble is external, it can function as a cue to others that an individual is in pain.[25] When the source is internal, the individual's pain functions as a cue that there is something physiologically wrong.

When the trouble is external and we know that the individual feels pain there, we immediately and naturally have found the cause of the pain. When the trouble is internal, we naturally assume that there is something amiss physiologically and quite often we have a fairly good idea what the cause is, e.g., one's child has eaten a box of candy and complains of a stomach ache. Though we cannot see the child's indigestion as it occurs inside his body, we would not normally panic and telephone the doctor—we know well enough, usually, what is going on and we know that the irregularity is neither serious nor permanent.

Thus some sort of physiological irregularity is expected and usually forthcoming when an individual is in pain. This is why it is part of our concept of pain and why I have included it among the aspects of pain. In almost all cases of pain such a factor will be present. But in cases where no such factor is found, after all the known observations and tests have been conducted, we often call the pain 'psychosomatic'. Here

---

[24] It is true that the neural processes are bodily processes but here I expressly exclude the neural processes and include any other bodily state or process which could normally figure as a source and/or sign of pain.

[25] The position of the person may also be a cue, e.g., if one slumps in one's chair, is doubled over or is flat on one's back. But here the distinction between physiological irregularity and *behavior* becomes unclear.

only the first two factors, the experiential and the neural, will be found (and perhaps one or the other or both of the fourth and fifth factors).

When we say that someone's pain is "only in the head", "mental" or "only in his mind", we are not (or should not be) denying that the individual in question actually feels pain, i.e., that the experiential aspect is present. What we are saying is that his pain lacks a physiological foundation, so far as we can tell. We assume that his pain must be the result of psychological, e.g., emotional factors. Or, if we call someone's physiological irregularity 'psychosomatic', we mean that its origin is psychological.

There are also cases of 'phantom pains' or 'phantom limbs'. For example, someone may feel that he has a pain in a limb which he no longer has. Hirst's remarks about such a case seem to me to be interesting:

....if one's foot is amputated one may afterwards feel that it is still there and feel pain in the non-existent toes; this is apparently because the nerves which previously connected toe to brain and which remain in the untouched part of the limb, are still sending impulses like those normally due to external excitations.[26]

These remarks of Hirst are reminiscent of those of Descartes in the sixth *Meditation* where he is trying to account for the fact that we are sometimes deceived by our senses even though we are creatures of a perfect God:

....if there is any cause which excites, not in the foot but in some part of the nerves which are extended between the foot and the brain, or even in the brain itself, the same movement which usually is produced when the foot is detrimentally affected, pain will be experienced as though it were in the foot, and the sense will thus naturally be deceived....[27]

The experience of pain had by the individual missing his limb can be qualitatively identical to the sensation of someone who has his limb. That is to say, the experiential aspect could be the same in these cases. This can happen because, no doubt, the neural, or at least cerebral, processes are the same. It is these two aspects, the experiential and the cerebral (usually *all* the neural ones), which alone constitute the sensation or experience proper.

However, one's experience of pain may be altered by one's awareness of some of the other aspects of the total situation, e.g., the sight of the

---

[26] Hirst, *Perception,* p. 148. However, Melzack suggests that the central nervous system itself is responsible; see "The Perception of Pain", p. 304.

[27] Descartes, *Works,* p. 198.

ugly wound or the realization that one is groaning audibly or wringing one's hands visibly. Thus, even though the sensation or experience itself may be said to be constituted by the cerebral and experiential aspects, it must be realized that the character of this experience (these aspects) is dependent on many factors, both physical and psychological.[28]

4. The behavioral aspects: This is an aspect of mind. That is, we often do refer primarily to someone's behavior when we speak of his mind, e.g., "He has a good mind". However, as noted before,[29] consciousness is a sufficient condition for the presence of a mind. If a person were conscious but unable to behave at all due to complete paralysis, we would still say that he had a mind. In fact, even if we allowed that a computer or a robot "behaved" in many ways as a man, we would not say that it had a mind if we knew that it was not capable of being conscious.[30]

Behavior, then, is often what we are referring to when we speak of someone's mind. But we usually are referring in such cases not merely to what the person did or is doing but also to what he is capable of doing. As stated before, Ryle's characterization of the mind *is* correct, so far as it goes. He goes wrong when he denies the existence of occurrent mental processes or states.

Behavior is a characteristic aspect of being in pain. X, who has a severe toothache, is more than likely to exhibit (manifest, show, display) his pain by a variety of bodily and behavioral phenomena. Some of these consist of things that happen to him (sweat pouring down his face, extremely heavy and rapid breathing, etc.) and some consist of things that he does (placing his hand to his cheek, opening his mouth as rarely as possible, refraining from dancing, etc.).[31]

But X is not just "more than likely" to express or display his pain in his behavior. If his pain is severe, he will likely find it impossible *not*

---

[28] Cf. above, p. 35.

[29] Cf. above, pp. 10-11.

[30] The notion of unconscious mental factors in us is not being denied. We would not call them 'unconscious' unless they were related in some way to *conscious* states or processes. We do not call trees or stones "unconscious"—we reserve this term for beings who at least *could* be conscious.

[31] Bodily aspects (what happens to X) shade into behavioral aspects (what X does) which shade into verbal aspects (what X utters or says). I shall not try to draw any hard and fast distinctions here in borderline cases because there *are* no such definite distinctions. We deal with the verbal aspects of being in pain after dealing with the behavioral aspects.

to express it. A pain is, after all, a kind of experience. It is an experience which we find slightly annoying, fairly distressing, or absolutely unbearable, depending on its severity and duration. It is a kind of experience which naturally causes certain behavior which everyone recognizes as pain-behavior.

The problem of other minds is currently exercising many philosophers. It can be stated in many different ways but the one which is most appropriate here is this: If pain is an experience, how can I know that another is in pain since all I can see is his behavior?

If the questioner makes it a condition of my knowing that X is in pain, that I *have* X's pain, then clearly nobody could ever know that another is in pain since each of us can have only his own pain.[32] Barring this necessity, however, it is relatively easy to see that someone else is in pain. This is because the pain-behavior is characteristically an *aspect* of being in pain. Pains naturally create characteristic behavioral patterns. We see this clearly in the case of infants and children.

It is natural and normal for babies to cry when stuck with a pin. It is as natural and normal for them to cry in such circumstances as it is for them to bleed. It takes years of training before a child learns to control his tears and to concentrate on eliminating his pain instead of complaining about it. Of course sudden, severe pains can bring tears to anyone's eyes, even adults. There are some things we can control and some that we cannot.

Evidence that behavior is a part of being in pain is readily available. Certain tribal societies conduct initiation rites which involve the inflicting of severe pains [33] on the adolescents. During these puberty rites, the individuals in pain are expected to suppress all the natural bodily, behavioral and verbal signs of their pains. They are expected to do so because it will ensure a high status in the society into which they are being formally initiated. It is extremely difficult to suppress these natural signs of pain—otherwise why would their suppression gain one social

---

[32] Of course you and I could have qualitatively identical pains, e.g., when we both suffer from identical physiological irregularities. But we could not have numerically identical pains.

[33] "All male Nuer are initiated from boyhood to manhood by a very severe operation (*gar*). Their brows are cut to the bone with a small knife, in six long cuts from ear to ear. The scars remain for life and it is said that marks can be detected on the skulls of dead men." E.E. Evans-Pritchard, *The Nuer*, Oxford, Clarendon, 1968, p. 249.

approval? One proves one's bravery and endurance by being able to experience pain without giving vent to it.[34]

Being in pain, then, involves not just an experience, neural processes and a physiological irregularity but also behavior. And it is not simply a coincidence that one in pain normally engages in pain-behavior. The experience is normally deemed undesirable and one wishes to avoid it, not to prolong it.[35] The Martians discussed earlier could not understand pain-behavior unless they could understand the experience of pain itself, because there is an intimate relationship between the experience and the behavior.

Isabel C. Hungerland discusses this peculiarly intimate relationship in "My Pains And Yours":

Not only does the sense of *being in pain* comprehend both inner and outer aspects, (feeling and behavior) but these are not synthetically related, not, that is, merely found empirically to be correlated, or causally connected. The reason is this. The notion of the unpleasant, the distressing is not just synthetically connected with the notion of avoidance behavior. (The notion of something which is extremely distressing but attracts every actual or possibly imagined person, angel, or God, is paradoxical.) Now, pain is distressing but it cannot, since it is in our bodies, be avoided, and the general features of pain-behavior are those which are appropriate to this sort of situation.[36]

Pain, in other words, is characteristically avoided, because it is characteristically considered to be undesirable. The reason it would be paradoxical if everyone were attracted to a distressing experience is that 'distressing' connotes anguish, affliction and pain—and an experience which attracted us would never have been *called* 'distressing' in the first place. Were we all masochists, I presume that what we now call 'pain' would have been called 'pleasure'. However, since pain is almost universally considered an undesirable experience, 'pain' connotes distress, suffering and displeasure.

It may be argued that no matter how severe the pain seems to be, judging from X's behavior, we can never be sure that he is really in

---

[34] "The transition from childhood to the status of an adult is effected...by ceremonies of initation...The boy...must undergo a long series of complicated rites, often of a highly painful or even dangerous nature....Their object is...to test his fortitude and qualifications for manhood...." George Peter Murdock, *Our Primitive Contemporaries*, New York, Macmillan, 1959, pp. 35-36.

[35] One may wallow in, or be contented with, pleasure whereas only masochists enjoy pain.

[36] *Loc. cit.*, pp. 121-22.

pain since we *do* have to judge from his behavior. After all, it may be said, X may be trying to deceive us or he may be a freak who shows all the signs of pain (even the neural ones) yet feels nothing.

Again the sceptic is making it a condition of being sure of X's pain that one *be* X. In fact, now it seems doubtful whether the sceptic even knows what it *means* for someone else to be in pain. Norman Malcolm, in "Wittgenstein's *Philosophical Investigations*", puts the point this way:

....if someone *always* had endless doubts about the genuineness of expressions of pain, it would mean that he was not using *any criterion* of another's being in pain. It would mean that he did not accept anything as an *expression* of pain. So what could it mean to say that he even had the *concept* of another's being in pain? It is senseless to suppose that he has this concept and yet always doubts.[37]

Someone who would never concede that another were in pain, then, could hardly be said to know what is *involved* in another person's being in pain. Quite simply, he has not realized that one needs no criterion or evidence to determine that oneself is in pain but one always needs it to determine that other people are in pain (since their bodies are not our bodies). He has not realized that pain-behavior does not merely coincide with, accompany, or attend upon pain. It is its natural product, just as the tears and cries of an infant are the natural product of pain. Pain is the sort of experience which *makes* us cry, double over, wince, shriek, writhe and/or faint, depending on its severity and on our desire and ability to suppress these natural manifestations of our pain.

In short, the normal situation is where the pain-behavior is proportionate to the severity of the pain—a slight ache makes us wince and an excruciating pain doubles us over. It is only in the abnormal cases —the puberty rites where there is much pain and little or no display; the case of the successful malingerer where there is great display and little or no pain— it is only in these cases that the behavior is inordinately disproportionate to the severity of the pain. The concept of pain, of one's own pain and of another's, is based on the normal cases of pain where the behavior is approximately proportionate to the pain's severity, not on the abnormal cases. The latter are parasitic on the former. One is familiar with the natural pain-behavior before one can suppress it or simulate it.

It is natural, then, to express one's pain by crying, groaning, or putting one's hand to where the pain is felt. We have shown this by our con-

[37] In *The Philosophy of Mind,* ed. by V.C. Chappell, p. 90.

sideration of the behavior of infants, adolescents during puberty rites in tribal societies and almost all people most of the time. Behavior is thus almost invariably one of the features of being in pain. It is usually as much a part of pain as are its neural, experiential and bodily features.

To say, with the Identity theorists, that sensations are nothing really but brain processes, is to ignore almost every known feature of the state of pain. How, for example, could we explain our pain-behavior in terms of the brain processes alone? We writhe because something *hurts* us. But how could *this* ever be expressed in terms of brain processes alone? And if it could not be so expressed, if sensations were *really* just brain processes and not a distressing experience, we could never understand pain-behavior.

5. The verbal aspect: This is not what X feels, not what is occurring in his brain, not what happens to him and not what he does—it is what he says. It may be objected that verbal behavior is just a species of behavior and, therefore, that it should not be classified separately.

I think, however, that the verbal aspect of pain merits separate classification. It is the one aspect of pain which is uniquely human. The other animals share with man all the other features of pain. Only man verbalizes his pain. I exclude here involuntary groans and include utterances such as "I'm in pain", "It hurts" and "The pain is getting much worse, Doctor".

This aspect, along with the first, second and fourth, is part of what it means to have a mind. Humans, unlike the rest of the animals, can "speak their minds". Ability to use language has been at least as important in human evolution as being able to use tools.[38] The development and transmission of culture—a distinctively human product—would have been impossible without language.

Language is not only distinctively human but it is also the most *expressive* of all our behavior. If someone doubts that X is in pain when X's tooth is visibly decayed, his face is flushed and he holds his cheek in his hand, X can *tell* him, can assure him, that he (X) is in pain. The doubter may then suggest that perhaps X is a deceiver, and perhaps all men are always deceivers. Were such a doubter actually to be found, outside of a philosophical essay, he would need reminding of the following point made by J.L. Austin in "Other Minds":

---

[38] In a sense, language itself is a tool—or tool kit—used for communication.

It is fundamental in talking (as in other matters) that we are entitled to trust others, except in so far as there is some concrete reason to distrust them. Believing persons, accepting testimony, is the, or one main, point of talking.[39]

There are borderline cases not only between the bodily and behavioral aspects, but between both of these and the verbal aspect. Is a scream something that happens to one, something that one does or something that one says? [40] I have stipulated that it is not something that one says but there are circumstances where the situation is not at all clear. For example, if someone were swimming alone in deep water and suddenly felt a severe, disabling pain, he might deliberately scream as loudly as he could for help. However, even though there may be borderline cases, it is useful to consider such utterances as "I am in pain" or "It hurts" where these are clearly voluntary utterances.

In cases where there is no visible physiological deformity and no pain-behavior, such utterances as "It hurts" or "I have a toothache" are usually decisive criteria of an individual's being in pain. The reason for this is not just because we are entitled to believe people unless we have specific reasons for doubting their word but because these expressions function, at least originally, as pain-behavior. Ludwig Wittgenstein makes this suggestion in his *Philosophical Investigations*:

....how does a human being learn the meaning of the names of sensations?—of the word "pain" for example. Here is one possibility: words are connected with the primitive, the natural, expressions of the sensations and used in their place. A child has hurt himself and he cries; and then adults talk to him and teach him exclamations and, later, sentences. They teach the child new pain-behavior....the verbal expression of pain replaces crying....[41]

Avowals of pain, then, can *prima facie* be trusted in the same way that the cries of a baby can be trusted. Both entitle us to infer that the individual is in pain, unless we have a specific reason to think that he may be shamming.

We argued earlier that one in pain needs, and could make use of, no evidence or criterion in order to be apprised of his own pain, whereas others need some evidence, sign or criterion. The reason for this is not that pains occur in ghostly minds but that our bodies, and brains, are independent. Wittgenstein suggests that Siamese twins might feel pain

---

[39] In Austin's *Philosophical Papers*, p. 50.
[40] Of course saying something may *constitute* doing something.
[41] *Op. cit.*, p. 89.

in the same place [42] and that this, presumably, would be a case of two people having the same pain. If this were so, it would vitiate my arguments regarding the privacy of pain.[43] Unless the twins were joined at the head and shared one brain, however, we would still speak of *two* pains. They may be qualitatively identical but they will not be numerically identical.[44]

People, then, do not need to make any observations in order to learn of their own pains. They just *have* a pain and then cry, exclaim or report, according to their age and to the circumstances. This is why avowals of pains bear a relationship to pains similar to that which bodily and behavioral signs of pain enjoy. We can apply what Ryle says of avowals to the case of pain:

> If a person says 'I feel bored', or 'I feel depressed' [or 'I feel pain'], we do not ask him for his evidence, or request him to make sure....The conversational avowal of moods [or *pains*] requires not acumen, but openness. It comes from the heart, not from the head. It is not discovery, but voluntary non-concealment.[45]

Children gradually move from crying and whining to saying "Ouch" and "It hurts". All of these are natural expressions of pain and it is only after we have seen this that we could successfully malinger.[46]

Avowals of pain are not characteristically uttered in the same tone as casual remarks about the weather or reports of someone else's pain. Usually they are uttered with a sense of urgency or concern, depending on the severity and/or duration of the pain. In many circumstances where pain occurs, the situation is such that we know what has caused or is causing the pain and we can see the bodily and behavioral signs of pain. In such cases we do not need to be *told* that the individual is in pain. When, for example, X is having his tooth filled at the dentist's, we should be surprised if he did not feel any pain. In general, when usual sources of pain are available for all to see, we do not need to be told that X will soon be, or already is, in pain. X may not be able to say anything because of the dental apparatus in his mouth, but in such

---

[42] *Ibid.*, p. 91.

[43] Cf. above, pp. 53 and 65 ff.

[44] Were two bodies shared by one head and brain, we would probably not speak of Siamese twins but of a person (or monster) with two bodies. We would probably continue to apply the dictum—one head, one brain, one person, one experience.

[45] *The Concept of Mind*, p. 102.

[46] In fact, the very notion of malingering involves deceiving *other people*. Thus the fear that everyone else may either constantly be deceiving us or may actually be automatons is groundless.

circumstances we will have enough evidence from the *surroundings* and from X's bodily and behavioral signs to be confident that he is in pain. When, however, X is in a position to tell us of his pain, he will often be able to express it to us in much greater detail and with greater clarity than could be conveyed by bodily and behavioral signs alone.

These, then, are the five aspects or features of being in pain— experiential, neural, bodily, behavioral and verbal.[47] The first two aspects are the *sine qua non* of pain [48] but some or all of the last three are characteristically present. Pain involves neither solely neural processes nor solely an experience. It is conceivable that someone could possess the latter of these features without the other but so far as we can tell this could never occur. Neither, of course, is pain solely behavior. Once it is learned that pain always has a neural component, any theory which identifies pain with actual and potential behavior is rendered untenable.

To say that one or the other of these five aspects is what pain (or the mind) *really* is, is to ignore the other four aspects. Identity theorists ignore all but the neural aspect. Cartesian Dualists deny all but the experiential aspect and Behaviorists deny all but the behavioral aspect.

As has been argued, these features of being in pain do not occur together by accident. They are all very intimately related. We could say of the state of being in pain what J.L. Austin says of the state of anger:

It seems fair to say that 'being angry' is in many respects like 'having mumps'. It is a description of a whole pattern of events, including occasion, symptoms, feeling and manifestation, and possibly other factors besides. It is as silly to ask 'What, really, *is* the anger itself?' as to attempt to fine down 'the disease' to some one chosen item ('the functional disorder').[49]

Being in pain is no more "really" neural processes than is being angry "really" a quickened heart beat or the weather "really" rain.

The desire to find the reality behind the appearances has been prevalent in philosophy since Thales conjectured that the world is really composed of water. But the world is no more really water than it is really atoms.[50] This is not to deny that objects may be constituted by atoms; it is to deny that objects are nothing but or are "really" atoms.

---

[47] Four of them (all but the bodily) constitute the aspects of mind.

[48] I have suggested that it is conceivable that an isolated brain might be said to have experiences; see above, pp. 20 ff. For arguments against its conceivability, see D. Murray, "Disembodied Brains", in *PAS,* January, 1970.

[49] "Other Minds", p. 77.

[50] And if pains were really neural processes, would they then be really atoms? (If so, then are they *not* really neural processes?)

To say that someone has a mind is to say that he is at least some of the time conscious, that he can act and speak reasonably intelligently and that he has a brain.[51] It is as misleading to say that the mind is simply the brain [52] as it would be to say that people are simply atoms or that the weather (during a storm) is simply rain. Identity theorists are captivated by a mistaken picture of explanation and reduction in the sciences.[53]

Suppose someone were to say that the weather during a particular storm is "simply heavy rains and strong winds". (We shall assume that he does not conclude from this that these in turn are *simply* atoms.) First we might point out that at least one crucial factor had been omitted, viz., the temperature. Then we might mention the 'chill factor' created by the winds. Once we had enumerated all the factors, we could say that all of them together constitute the weather on this particular occasion. Unless someone thinks that the weather is something *over and above* the general condition of the atmosphere with respect to all these factors, there would seem to be no point in saying that the weather is "simply" or "nothing but" these factors. Providing it is understood that the elements constitute a whole or a totality called 'the weather', it is not misleading to say that the weather *is* all its factors. And so long as it is understood that all the elements of a person constitute a *person,* it is not misleading to say that a person *is* all the components which make him up. It would, of course, be misleading simply to identify a person with the atoms (or electrons and protons) which constitute him because this omits reference to the molecules which the atoms constitute, the organs which the molecules constitute, and so on—not to mention the person's hopes and dreams, etc.

It is important to note that there are many different types of mental states, processes and activities. Just as there is no thing or substance called 'the mind', so there is no simple definition or essence of mind or mentality. In effect, Descartes claimed that consciousness was the essence of mind and Ryle claimed that intelligent behavior was its essence. In fact, however, mental states and activities form a kind of continuum or spectrum which includes such variegated phenomena as pain, anxiety, anger, day dreaming, calculating and playing chess. Is there any feature

---

[51] Notice that we often use such locutions as: "Use your head"; "He hasn't a brain in his head"; "He's very brainy"; "Does she have any brains?"; etc.

[52] Cf. above, p. 27.

[53] Cf. above, pp. 47 ff.

which is shared by all of these states or activities which makes them *mental*? I do not think there is. What we have here is, in Wittgenstein's words, a case of "family resemblances".

Phenomena which we call 'mental' form a spectrum or family each of whose members share one or another feature with some of the other members but do not share any single trait with the other members which makes us call them all 'mental'. Not all of them need involve intelligent behavior or awareness of their existence or privileged access or intentionality; but all of them, it would seem, will have one or another of these features, or perhaps some other trait which prompts us to classify them as 'mental'. I have already outlined some of the facts about pain which prompt us to call it a mental phenomenon. These features of pain constitute various criteria of the mentality of pain, though not necessarily criteria of the mentality of other sorts of mental processes, states or activities.

What I am suggesting is that the word 'mental' can mean something quite different when applied to pain than when applied to such phenomena as chess playing, being angry or remembering. What makes a pain mental, then, need not be what makes chess playing mental. Just as the word 'real' may mean something different when applied to different things (e.g., diamonds, ducks, persons or murders) and the word 'good' may mean different things (e.g., a good time, typewriter, child or good weather), so the word 'mental' need not always mean the same thing.

For example, it is obvious that chess playing normally involves more or less intelligent behavior. Thus chess playing might not have been as good an example to use in a discussion of the mind-brain problem, though it might have been acceptable in a discussion of the mind-body problem. Sensations need not figure largely in a game of chess whereas pain itself is a sensation. Identity theorists have focussed their attention on sensations; thus it was deemed advisable to use the example of pain in this discussion rather than some other sort of mental state or process.

It may be objected that pain is not a mental phenomenon at all. Any animal, it may be argued, can feel pain, but animals cannot reason, daydream or doubt, i.e., they cannot execute typical mental activities. What, it may be asked, is *mental* about pain?

Such a question presupposes a view of the mind or mentality not unlike that of Ryle's. We saw that his concept of mentality is too narrow. In effect, he denies that there *are* any mental states. He recognizes only mental *activities,* i.e., behavior which embodies some degree of intelligence and/or some quality of character. It was shown, however, that in a perfectly acceptable sense of 'mental', sensations are mental because they cannot sensibly be said to exist except as aspects of consciousness. That is, sensations are *conscious* states. To have a pain is to undergo a certain kind of sensation and it is self-contradictory to say that some organism is undergoing a sensation of which it is totally unaware. We are not claiming that *all* mental processes or activities are necessarily conscious; for example, a person may be undergoing anxiety and not notice it in a given instance, but this does not seem to be possible in the case of pain. What is being claimed is that all states which could not be sensibly said to exist unless their possessor were conscious of them are *ipso facto* mental states. And pain is an eminently good example of such a state.

All states of which one must be aware when they occur are mental states but it may not be the case that all mental states, e.g., anxiety, are states of which one must be aware.

We also noted other features of pain which prompt us to classify it as mental. There is the asymmetry in the way the one in pain 'learns' of his pain and the way in which others do.[54] This arises because the one in pain needs no criteria or evidence to determine that he is in pain, whereas others do require this.[55] Pains are contingently private [56] in the sense that one can suppress their natural signs and/or abstain from telling others about one's pain. Thus some instances of pain may involve no behavior whatsoever.[57] They are also logically 'private' in the sense that any pain which one feels is *ipso facto* one's own pain, i.e., my pain and yours may be qualitatively identical but they could not be numerically identical.[58] Finally, there is the feature called 'intentionality' which seems to be common to all experiences.[59]

These are the features of pains which lead us to classify them as mental phenomena. Whatever sort of 'stuff' sensations are composed of, these features will continue to lead us to classify them as mental states.[60]

We have argued that the mind-brain problem can be discussed independently of the mind-body problem. It may be objected that sensations

---

[54] Cf. above, p. 88 and p. 126.
[55] Cf. above, p. 88, p. 126, p. 68 and pp. 139-40.
[56] Cf. above, p. 65.
[57] Cf. above, p. 135.
[58] Cf. above, p. 54.
[59] Cf. above, pp. 128 ff.
[60] Since this was written, a paper called "Materialism and Immaterialism" by M.J. Budd has come to my attention in which this same point is made; *PAS,* March 23, 1970, p. 217.

involve behavior and that therefore the mind-body problem is unavoidable, or perhaps that it is indistinguishable from the mind-brain problem. However, even if pains usually and naturally involve pain-behavior, this does not alter the fact that pains are sensations—conscious states— or that they are invariably correlated with brain processes. I see no reason why we cannot discuss the relationship between sensations and brain processes independently of the relationship between mental states and behavior. To deny this would be like denying that we could discuss the question of the relationship between a cloud and the mass of particles of which it is composed independently of the question whether clouds sometimes move very quickly. (It must be admitted, however, that the *concept* of pain does seem to involve behavior whereas the concept of brain processes does not seem to do so.)

As for the objection that since even a dog can feel pain, there may be nothing mental about pain, I would add this: If there is any sense at all to the notion that the severed heads of monkeys and dogs could still be conscious, and I submit that there is, then I see no difficulty in the notion that monkeys and dogs which are *intact* are sometimes conscious. If the arguments designed to show that consciousness is one criterion of mentality are acceptable, then we must admit that wherever there are conscious states, there is mentality. Of course if dogs do not doubt or reason, they cannot be said to have minds in the same sense or to the same degree as persons. But anyone who wants to argue that there is nothing *mental* about pains should feel no qualms whatever about conducting such experiments on severed heads as those outlined earlier.

Why, after all, should we not deliberately inflict grievous bodily damage, not only on dogs and monkeys, but even on human subjects? We might learn a lot about pain by permitting any sort of experiment to occur. For example, we could administer electric shocks to human subjects and determine exactly how long they can endure the pain before they lose consciousness or before they die. Or as has already been done, we could place different chemicals on the eyes of rabbits and determine how long it takes before they rot away completely. This should help us develop more effective nerve and chemical weapons.

What is the difference between conducting such experiments on pieces of steel and conducting them on living organisms? Is it not because the latter are sentient and the former are not? Living organisms can feel pain and must to some degree be conscious of this state. It may be difficult to determine exactly where in the evolutionary scale sentience and consciousness begin, but granted that the expression 'unconscious pain' is self-contradictory, we can only assume that where there is pain, there is some degree of consciousness.

It is not the case that if steel were conscious it would feel the pain which we are inflicting upon it. The point is that it is senseless to speak of *pain* where there is no *awareness*. And this is why we are disquieted at the thought of inflicting needless pain on any organisms in the name of 'science' or for any other purpose. It is not so much the bodily damage that we deplore as the pain that is caused; we are, at least in part, sickened at the thought of an unanaesthetized rabbit's eye rotting away because we know that it cannot be occurring without some pain being involved. And our disquietude would make no sense whatsoever if there were not some degree of consciousness involved here. If rabbits were not at all conscious, then they could feel (be aware of) no pain—they would have no pain—and Descartes would have been right in saying that other animals besides man are automata. If animals were automata, they would no more feel pain than a piece of steel feels pain.

The dawning of sentience in the scale of evolution marks the dawning of one criterion of mentality, i.e., awareness. The beginning of intelligent behavior, behavior which involves learning, planning, imagination, etc., marks the appearance of another criterion of mentality. Whether there is any connection between these criteria is not relevant here. What is important to see is that intelligent behavior is not the only criterion of mentality. Wherever there are conscious states, whether behavior is involved or not, neural firings will be correlated with them whether they are found in men, monkeys, dogs or rabbits. I have restricted my discussion to pain in humans because there can be no doubt that we sometimes feel pain whereas in the case of some of the animals (e.g., earthworms) there are some who might doubt that they feel pain. Also, of course, some of the facts about pain outlined above may not apply to other animals, e.g., the ability to abstain from exhibiting pain or from telling others about it.

In many cases, of course, the word 'mental' may refer primarily to behavior and only secondarily, if at all, to sensations or experiences. For example, to say that someone is a 'mental case' usually means that he acts oddly. To say that someone is 'mentally disturbed', on the other hand, draws attention more to his experiences and trains of thought than to his overt behavior. We shall conclude this discussion of the various types of mental states and activities with a series of statements, some of which may refer primarily to experiences and others of which may refer mainly to activities. Reflection on these statements should satisfy us that having a mind does not consist solely in behaving intelligently.

"I cannot begin to tell you the mental torment that I have suffered."

"He has a good mind."

"Poor fellow—he is losing his mind."

"He seems to be having a bad trip." (Said of one 'on' LSD).

"Our cat is terrified of mice."

"To prove that two heads are better than one, Russian scientists transplanted the head of one dog onto the body of another dog—I wonder if the dogs were convinced."

"We should cease these experiments immediately; if these animals were not dumb, they would soon tell us that they are in pain."

"As I push the needle in further, tell me how it feels."

CHAPTER SEVEN

KINDS OF PAINS AND KINDS OF LANGUAGES

We have been discussing pain as though there were only one type of pain which may be felt in a tooth, a foot, the head or the stomach. There are, in fact, many different kinds and degrees of pain.

Pains may be throbbing, stabbing, piercing, stinging, gnawing, dull or sharp. They may be localized in one area of the body or they may pervade one's entire body. Pains may vary in degree from being annoying to being troublesome, bothersome, excruciating or intolerable. Many different sorts of things can be said about pain which could not be translated without loss into talk about neural firings.[1] In all cases of pain, no doubt, there will be neural processes involved, but the pain experienced is a complex of such factors.

Experiences of pain may differ in all of the above ways. The differences are due partly to the nature of the source of the pain and partly to the psychological factors involved.[2] Thus, a burn usually feels different from a scrape and both of these differ from a cut. One may be bitten, kicked, stabbed, pricked, slapped, sat upon, knocked over, battered, whipped, caned or strangled. What one feels in each of these cases, i.e., one's experience of pain, may differ considerably from what one feels in the other cases, depending upon one's unique past history and the circumstances of the particular situation.

The kind and degree of pain one feels, then, is a function of the nature of the internal and/or external source of the pain and of the psychological factors. Thus reference to neural processes alone could not suffice as an elucidation of what it *feels like* to have a given type of pain. When describing our pains to others, we characteristically refer to the situation

---

[1] Cf. above, p. 99.
[2] Cf. above, p. 35.

and the events which we suppose to be the source of our pain. For example, if asked what sort of stomach ache we had, we might say that it was a dull, heavy ache like the one we sometimes get from eating too quickly. If asked to specify a headache, we might say that it felt as though someone had tightened a metal band around our head. Again, an ache in the back may be described as one which feels as though it had been crushed by a huge truck.

Our vocabulary for describing the quantity and quality of our pains is not as sparse as it is often said to be. We usually succeed in conveying to others what we are feeling. Very often they will register their understanding by shaking their head knowingly and/or by saying something like this: "Oh yes—I've been hit by that same truck" or "Oh yes—and I'll bet it hurts worse at night than during the day".

It is certain that each of these different sorts and degrees of pain will have a neural aspect. It is even highly probable that the neural features of each pain will differ according to the type and degree of that pain. But in each case the state of being in pain will be an ensemble or composition of not only the neural factors but the others outlined above as well. Even if the experience were composed exclusively of neural processes, as the Identity theorists would have it, it is not at all *reducible* to these neural processes, as we have seen.

J.J.C. Smart seems to think that we could say the same things about our sensations as we now do by employing a physicalistic language, i.e., employing a language which makes reference to none but physical objects and processes.[3] We have argued that it would neither be helpful nor possible to talk about our sensations as though they were brain processes.[4]

It may be true, as Smart claims, that "terms such as 'wax' and 'wane'.... are equally applicable to brain processes and purely 'psychic' processes".[5] But it is also true that some terms applicable to our experiences are not equally applicable to brain processes. The pain that I would describe as 'throbbing' *feels* a certain way to me. Though the neurons which are correlated with this pain may be firing intermittently and in exact proportion to my experience, we would not say that the neurons or the firings were "throbbing". And if we would say this, it would not convey what "It is throbbing" said of our *pain* conveys.

[3] Cf. above, p. 45.
[4] Cf. above, pp. 46 ff.
[5] *Scientific Realism,* p. 103.

And what of 'gnawing' pains? In what sense could neural firings 'gnaw'? Or 'stab'? Or 'pierce'? And pains can have an intensity which no amount of talk about the firing of neurons could convey. How, for example, would we describe the firing of neurons as 'exquisite'? (Except, perhaps, to say that they (it) *feel(s)* exquisite!)

It is dubious whether even 'wax' and 'wane' mean the same when applied to brain processes as when applied to pains. If a physical thing waxes it increases in size, quantity, volume or intensity (e.g., a light bulb getting brighter and at the same time, somehow, growing in size). A pain may become more intense but how can it be sensibly said that an *experience* increases in size, quantity or volume? Also, the intensity of pain does not mean the same *sort* of thing as the intensity of neural firings. The firings may increase in number, occur more rapidly and become more widespread but none of these, nor all together, can convey what "It hurts more" conveys.[6]

As noted before, scientists have always found it necessary to employ a dualistic terminology when they describe their experiments on pain and other experiences, in *humans*. It may be asked why we find it necessary to experiment with pain in humans at all. "Why not work with other animals, if we must conduct such experiments?" it may be asked. Why, in fact, should we bother to conduct any experiments on pain? Should we not just try to eliminate it?

Ronald Melzack, who has conducted experiments on humans, expresses clearly the importance of pains:

[Pain] warns us that something biologically harmful is happening. The occasional reports of people who are born without the ability to feel pain provide convincing testimony on the value of pain. Such a person sustains extensive burns and bruises during childhood, frequently bites deep into his tongue while chewing food, and learns only with difficulty to avoid inflicting severe wounds on himself.[7]

Scientists do not study pain in order to eliminate it, therefore, because we know that it is crucial for survival. They study it in order to find out its neurological and psychological foundation and its neural correlates so that it will be better understood and thus predictable and controllable. We want to know enough about pain so that we can give an infant who is unable to feel it the ability to do so and can eliminate, from those

---

[6] A brain process might exhibit characteristics *corresponding* to those of the sensation, but to acknowledge this is, in effect, to deny an *identity* between the brain and mental processes.

[7] "The Perception of Pain", p. 299.

suffering unnecessarily, the ability to feel it. And in order to gain this knowledge it seems likely that we shall at some point have to experiment on human beings.

Of course we should keep experiments on pain to a minimum, among all animal species (though a consistent Identity theorist could claim that pains were nothing *really* but a certain type of neural process). Experiments on humans seem necessary because our brains do differ in some respects from those of other animals but, more importantly, humans can *tell* us what sort of experience they are undergoing whereas other animals can only act accordingly. This is not to say that there is any ground for doubt that other animals besides man feel pain. The behavioral repertoire of many animals more than compensates for their lack of speech. Humans, however, can tell an experimenter exactly what sort of pain they are feeling and where they are feeling it—from a barely perceptible pain to an excruciating pain. Animal experiments would be greatly facilitated if they could speak to us—but then we might be reluctant to pursue some of these experiments.

We need, in other words, the *introspective* method in the study of pain in humans. We might be able to get along without it, i.e., without using first person pain reports (avowals) but this would make the study of pain much more difficult and prolonged. There are some things which it would be difficult, if not impossible, to discover without it, e.g., the precise second when an individual begins to feel pain as opposed, say, to a state of discomfort. Also, it would be impossible to convey the exact quality of our pain without resorting to language.

I am not advocating a return to the Introspective method as a substitute for what scientists are now doing. I have in mind precisely the sorts of experiments which Penfield describes [8] and which Melzack reports.[9] Science should and does utilize all the possible sources of data. Exactly what sort of experience a person is undergoing—a detailed and comprehensive description of his experience—is discoverable only by his testimony. And this will in all probability always be the case. The sort of knowledge suggested earlier regarding detailed correlations between all our feelings and all our thoughts [10] on the one hand, and our neural processes on the other, is "light-years" away in terms of present day research.

---

[8] Cf. above, pp. 39-40.
[9] Cf. above, p. 35.
[10] Cf. above, pp. 21 and 90 ff.

In fact, we may never attain such a level of understanding. There are 10-15 billion nerve cells in the human brain and these interact in different ways. Recent studies on memory and its neurological aspects give some indication of the complexity of the problems involved. For example, with regard to how many memories might be accumulated throughout a lifetime, Ralph W. Gerard, a professor of neurophysiology, says the following:

> Some tests of perception suggest that each tenth of a second is a single "frame" of experience for the human brain. In that tenth of a second it can receive perhaps a thousand units of information, called bits. In 70 years, not allowing for any reception during sleep, some 15 trillion bits might pour into the brain and perhaps be stored there. Since this number is 1,000 times larger than the total of nerve cells, the problem of storage is not exactly simple.[11]

The vast majority of these memories will never reach consciousness, unless we are hypnotized, drugged or our brain is probed. But the neurophysiologists are confirming what Freud said, viz., that we retain in memory all that we have ever experienced—and more besides. Gerard says that "the most intriguing problem about memory....is not the existence but the tremendous specificity of recalls".[12] The mechanisms of memory are now under intensive investigation. Memory, says Gerard, must be explicable in terms of the temporal and spatial patterns of the firing of impulses from the neurons. He says that we can assume quite reasonably "that a given memory is not represented by one specific local change but by a pattern of many changed loci—a pattern with sufficient redundancy so that if part of it is destroyed the rest will still suffice to represent the memory".[13] This suggests that each memory involves thousands of neural firings. We might ask the Identity theorists how many of these a memory is 'identical' with—should they identify memories with their neural correlates.

In the course of this article, Gerard mentions the finds of scientists concerning electrical stimulation of the brain and its effects in *consciousness,* e.g., different sorts of *sensations* and "the conscious recall of quite specific events from an individual's past".[14] It will be noted that Gerard employs a dualistic terminology—one for neural and other physical processes and one for experiences. In fact, all scientists doing

---

[11] "What is Memory?", in *Psychobiology,* p. 126.

[12] *Ibid.,* p. 127.

[13] *Ibid.,* pp. 130-31.

[14] *Ibid.,* p. 130.

such research employ a dualistic terminology. How *else* could they refer to experiences and how else could they convey what a dualistic terminology conveys? What an individual experiences and what we could see (the neural processes) are different features of the total phenomenon. This is why the Identity Theory has not been accepted, or even taken seriously, by neurophysiologists.

Philosophers may insist that all questions are physical questions [15] but the scientists who are actually studying the phenomena in question—experiences and their neurological basis and correlates—cannot avoid stating their experiments and findings in terms of both physical and psychological states and processes. This is not surprising, since man is a psychophysical organism. If we wish to investigate experiences in order to determine their neural correlates, we cannot begin by denying the existence of these experiences. Yet this is what P.K. Feyerabend would have us do.

In "Materialism And The Mind-Body Problem", he suggests that "if you want to find out whether there *are* pains, thoughts, feelings in the sense indicated by the common usage of these words, then you must become....a materialist".[16] In another article, "Comment: Mental Events and the Brain", he says that monists (Identity theorists) should try to develop their theory without using the existing terminology. But in the first paper cited, he suggests that materialism might be able to provide us with synonyms, presumably for the terms we now use to refer to and describe sensations, etc.

Now either we want to eliminate terms referring to psychic processes or we do not. If for some reason we do, we would not want to reinstate synonyms for them because all we would have done is invent new words to do the jobs the old ones did. And if the old terms generated confusion, so will the new ones. If we do not feel the need to eliminate terms for psychological states, then we cannot be materialistic monists, i.e., Identity theorists.

Feyerabend wants to develop a language which is "fully testable".[17] By this, presumably, he means that our language should make reference to none but physical, i.e., *intersubjectively testable* entities, states and processes. But he then acknowledges the need for *introspection*. It is introspection with a twist, however.

[15] Cf. above, pp. 48 ff.
[16] *Op. cit.*, p. 53.
[17] "Materialism", p. 55.

He begins by warning us not to presuppose that "things inside the skin are very different from what goes on outside".[18] Since it is difficult to observe the micro-processes in the brain of a living organism, we need only consider the following facts:

....the living brain is *already connected with a most sensitive instrument*— the living human organism. Observation of the reactions of this organism, *introspection* [emphasis mine] included, may therefore be much more reliable sources of information concerning the living brain than any other "more direct" method. Using a suitable identification-hypothesis one might even be able to say that introspection leads to a *direct observation* of an otherwise quite inaccessible and very complex process in the brain.[19]

As if Wilder Penfield had only to look a little more closely at his patient's brain in order to see what the patient sees (or feels)! And this is just the point. The experiences which the subject is undergoing could not be *observed* by anyone, not even the subject himself (though he has the experience). Experiences are not that sort of thing.[20]

Feyerabend suggests that adopting an Identity Theory might enable us to say that the subject has "direct observation" of processes in his brain. But adoption of any theory at all will not change introspection into observation, direct or otherwise (whatever Feyerabend may mean by "direct observation"). We do not *feel* pain in the same sense as we *feel* a piece of paper. Seeing events in our "mind's eye", as we relive past experiences, is not the same thing as seeing a table in front of us. The former are 'private' in the senses outlined earlier.[21] You could not feel my pain, let alone observe it or relive my past experiences, let alone observe them.

The fact is, we do not observe our brain processes when we observe something in our mind's eye nor do we feel brain processes when we feel pain. Nor do we see atoms or molecules when we see a table, though tables may, indeed, be composed of molecules which are in turn composed of atoms. Thus the "direct observation" which Feyerabend claims that individuals have of processes in their brain has nothing in common with ordinary introspection. Brain processes are not the subject of first person psychological statements.

[18] *Ibid.*, p. 55.

[19] *Ibid.*

[20] If it is asked what sort of thing they *are*, all one can do is remind the inquirer of what happens to him when he is in pain or undergoes other experiences. Experiences cannot be defined in intersubjective terms because they are *not* intersubjective.

[21] Cf. above, pp. 65 ff., *passim*.

Feyerabend sees the value of, and perhaps even the need for, introspection. We too have insisted that it is an indispensable method. But it is not a method for revealing physical properties of brain processes. Introspective reports may give us clues to the nature of the neural basis and correlates of experiences but they do not constitute perceptions of anything at all, least of all neural processes.

Finally, we shall consider the suggestion of Richard Rorty that "What people now call 'sensations' are identical with certain brain-processes".[22] He calls his version of the Identity Theory a "disappearance" version, viz., he holds that "sensations may be to the future progress of psychophysiology as demons are to modern science. Just as we now want to deny that there are demons, future science may want to deny that there are sensations".[23]

Rorty would have us say "My C-fibers are firing" instead of "I'm in pain".[24] He says that if we did this, it would simplify matters, at least for "the science of the future".[25] And he adds that "Elimination of the referring use of....'sensation' from our language would leave our ability to describe and predict undiminished".[26]

Rorty admits that this elimination would be very *impractical*.[27] But if there really *are* no sensations, as there really are no *demons,* it is curious that it would be impractical to eliminate sensation-talk in favor of brain process-talk. Few people feel any loss accruing from the cessation of demon-talk. Why could we not just stop talking in terms of 'pains' or 'sensations' and begin talking in terms of "C-fibers firing"? [28]

As suggested before,[29] we rarely say such things as "I'm in pain" to

[22] "Mind-Body Identity, Privacy, and Categories", in *Philosophy of Mind,* ed. by Stuart Hampshire, p. 35.

[23] *Ibid.,* p. 37.

[24] *Ibid.*

[25] *Ibid.* J.J.C. Smart also has linguistic revision in mind. In *Philosophy and Scientific Realism,* he conceives of one of the tasks of philosophy "as the rational reconstruction of language so as to provide a medium for the expression of total science", p. 2.

[26] Rorty, *op. cit.,* p. 39.

[27] *Ibid.*

[28] Presumably we could stop talking about tables and chairs also. Rorty cites approvingly the (by now familiar) statement: "What people call 'tables' are *nothing but* clouds of molecules". He adds, curiously, that "This is a table" may still be a true statement, though he does seriously suggest that the word 'table' is theoretically eliminable; *Ibid.,* pp. 40-41, emphasis mine.

[29] Cf. above, pp. 122 and 140-41.

each other. As children we cry and say "ouch", as adults we say "damn it" and can describe our pain in detail, should the need arise. When a child cries, should we inform him that what is really happening is that his C-fibers are firing? Or should we wait until he says "ouch" or even "damn it"? At what point in his life are we supposed to inform him that he isn't *really* undergoing a pain and that what is really happening is that his C-fibers are firing?

Here Rorty would insist that people will still feel as they have always felt when they (thought they) were in pain except that when the Identity Theory is universally adopted they will employ a monistic (materialistic) language which does not mention any psychological or private entities or processes. Rorty says that there would be no loss of meaning by replacing talk about pains with talk about C-fibers firing because then we would be reporting what is *actually going on* instead of talking about invisible entities.[30]

One of the difficulties with Rorty's suggestion is that people will still want to tell each other how they *feel*—how it *feels* when one's C-fibers are firing in this particular manner.[31] That is, we will still want (need) to exhibit and describe what is hurting us. Will Rorty and the other Identity theorists now insist that our distress or misery is itself really only brain processes? If they are consistent, it would seem that they must do this.

It is evident that we will continue to convey to others exactly what we now do. To say "My C-fibers are firing" will come to mean precisely what "I'm in pain" now means. None of the facts about a person's privacy or epistemologically 'privileged' access (in regard to his own experiences) or the experiences he undergoes will have changed. There will continue to be a radical epistemological difference between self-knowledge and knowledge of others. No philosophical theory can alter these facts. Thus no advantage would accrue from speaking "Materialese" rather than in terms of pains. It is, in short, impossible to devise an artificial language which will both refer only to what is intersubjectively testable and yet convey what we need to convey to each other.

---

[30] Rorty, *op. cit.*, p. 37.

[31] There is, of course, the difficulty about the new need to discover which C-fibers are firing and exactly *how* they are firing, if we are to convey anything significant about our pains (C-fibers firing). This will, presumably, require all people to *become neurophysiologists* before they could describe their pains (C-fibers firing) accurately! As we indicated earlier, practicing neurophysiologists recognize the need for a dualistic terminology—only philosophers want to eliminate it.

In addition to these difficulties, there is the one regarding an infinite regress in reduction. If the Identity theorists are right in claiming that sensations are nothing really but brain processes, in the sense in which tables are nothing really but clouds of molecules,[32] then surely these brain processes are not ultimately real. They too must be composed of, and therefore reducible to, molecules. And the molecules, in turn, are not ultimate particles—they must be nothing really but....*ad infinitum.* In brief, once we begin reducing things to their constituents, we cannot, in honesty, stop until we have reached something that has no constituents at all, i.e., until we have reduced everything to nothing at all.[33]

'Materialese' would not just be impractical or entirely useless for everyday purposes. It would make much neurophysiological investigation impossible. Suppose, for example, that we wanted to study the effects of LSD. We would give the volunteers the required dose and we would then wait to see what happens. If the volunteers were consistent Identity theorists, they would speak only 'Materialese'. How, then, could we learn about their experiences while under the influence of LSD? No amount of talk about neural firings could ever convey "the experience", as it is called. (It is apparently almost impossible to convey in a *natural* language.)

All these remarks apply to the experience of pain and to all other experiences. The Identity Theory is useless as a scientific hypothesis and any language based upon it would be useless for ordinary purposes. Reductionistic, materialistic theories have been known since pre-Socratic times. There is virtually nothing new about the Identity Theory which could account for its increasing acceptance among philosophers. It is an ancient metaphysical dream, not a workable scientific hypothesis.

---

[32] Which in turn are nothing really but atoms which in turn are nothing but....?
[33] Cf. above, p. 141.

CONCLUSION

Many philosophers are interpreting the recently discovered correlations between certain mental and neural processes as confirmation of the materialist, reductionist theory that experiences are nothing really but particles in space. We have shown, however, that the increasingly popular Identity Theory of the mind is a dead end both in philosophy and for the purposes of science.

The Identity Theory is neither more nor less testable than various Dualistic theories, e.g., the Correlation Hypothesis. Therefore it is not a fruitful scientific theory. Neither is it an adequate philosophical analysis of mental processes or the mind. Using the example of being in pain, the Identity Theory was exposed for the reductionist theory that it is. It was shown where it would lead if we followed it to its logical conclusion, viz., to nothing.[34] Since atoms are constituted by lesser particles and these themselves will always be subject to further analysis and factorization, the logic of the Identity Theory is such that we could never know what pains (or anything else) are 'really'. This indicates the folly of trying to identify an item or its 'real nature' with its ultimate constituents.[35]

An alternative theory of pain and of the mind has been proposed in this discussion. It was argued that the mind is not the 'real' person— an incorporeal thinking substance which interacts with a body to which it is contingently attached. Nor does having a mind consist solely in being able and prone to do certain sorts of things. The dispositional account of the mind leaves out occurrent mental states or processes.

Using the example of being in pain as such a mental state, the inadequacies of the Identity Theory were demonstrated. The mind is

[34] Cf. above, p. 157.
[35] Cf. above, n. 50 p. 141.

no more simply the brain than it is an incorporeal substance or a set of dispositions to behave in certain ways.

An alternative theory of mental states and the mind has been proposed and defended. A Multi-Aspect Theory of these phenomena was shown to be compatible both with the ordinary conceptions of persons with their 'privacy' and 'privileged access' and also with the methods and findings of the most recent neurophysiological investigations.

The five aspects of pain are the experiential, neural, bodily, behavioral and verbal. The four aspects of mind are the experiential, neural, behavioral and verbal.

Having a mind, then, normally consists in having experiences and a healthy, functioning brain. It also normally consists in being able to do certain sorts of things, e.g., to use a sophisticated means of communication. We acknowledged that people might be said to have a mind who lacked one or the other of these features but that the concept of mind which we have is based on the characteristic possession of all of them.

It was shown that a dualistic terminology is inevitable, considering that persons are psychophysical beings. No metaphysical theory can legislate away sensations or experiences. Even if pains are composed of neural processes, it was demonstrated in many ways that they cannot be reduced to neural processes. If things are reducible to their constituents, then there is nothing in the world 'really' but the ultimate particles. But once it is realized that the parts of an item do indeed constitute that item, the temptation to identify things with their smallest constituents, in the sense of reducing them to these, should disappear. Wholes can no more be identified with their parts in that sense than parts can be identified with what they constitute.

The Multi-Aspect Theory of the mind is, in my opinion, the most plausible one. It is eclectic in that it embodies the insights of alternative views. For example, Descartes was right inasmuch as the experiential aspect of the mind is the most important one. Nothing would be said to have a mind which lacked this feature. (This is one reason why we are reluctant to say that a *machine* could have a mind.)

Ryle was correct insofar as intelligence or intelligent conduct (including speech) is an extremely important and extensive feature of the mind. That is, many mental predicates do refer to our behavior and dispositions to behave, and not to our experiences.

Identity theorists emphasize correctly the close connection between brain processes and experiences. It seems evident that there is no mental 'stuff' called the mind in which mental processes occur. However, in contrast with the Identity theorists, we showed what it actually means to call a process 'mental' [36] and we argued that the reality of the experiential aspects of being in pain cannot be denied without absurdity. We also argued that persons are living organisms and are not just very complicated physico-chemical mechanisms.

Finally, we have shown that simple definitions of pain or the mind must be replaced by detailed descriptions and analysis. We saw what was involved in the case of being in pain. There is more occurring than any of the hitherto proposed theories of the mind has admitted.

There are several aspects of mind and each—or certain combinations of them—may be regarded as a criterion of our ascription of a mind to any entity or being. This enables us to see our way to a solution, or dissolution, of the puzzles about the possibility of ascribing a mind to machines, certain of the animals, or perhaps a being from another planet.

Lastly, and most importantly, the Multi-Aspect Theory of mind enables us to do justice to the facts about ourselves, something which narrower theories of the mind cannot do.

---

[36] Cf. above, p. 88.

REFERENCES CITED

Adrian, E.D. *The Physical Background of Perception*. Oxford: Clarendon, 1947.

Armstrong, D.M. *A Materialist Theory of the Mind*. London: Routledge and Kegan Paul, 1968.

Austin, J.L. "Other Minds" in *Philosophical Papers*, J.L. Austin, pp. 44-84. Oxford: Clarendon, 1962.

Ayer, A.J. "Mind and Matter" (A Review of D.M. Armstrong's *A Materialist Theory of the Mind*, London: Routledge and Kegan Paul, 1968) in *New States-man*, April 5, 1968, 453-54.

Brain, W. Russell. *Mind, Perception and Science*. Oxford: Blackwell, 1951.

Budd, M.J. "Materialism and Immaterialism", *Proceedings of the Aristotelian Society* (unbound paper), March, 1970, 197-217.

Calder, Nigel. "Notes and Comments: Experiments with Live Monkey Brains", *New Scientist*, 27 (1965), 319.

Descartes, René. *The Philosophical Works of Descartes*, 2 Vols., eds. E.S. Haldane and G.R.T. Ross. London: Cambridge University Press, 1967.

Dostoevsky, Fyodor. *The Devils* (translated by D. Magarshack). Middlesex: Penguin, 1967.

Eccles, John C. (ed.). *Brain and Conscious Experience*. New York: Springer-Verlag, 1965.

Evans-Pritchard, E.E. *The Nuer*. Oxford: Clarendon, 1968.

Feigl, Herbert. "The 'Mental' and the 'Physical'" in *Minnesota Studies in the Philosophy of Science*, Vol. 2, eds. H. Feigl, M. Scriven and G. Maxwell, pp. 370-497. Minneapolis: University of Minnesota, 1958.

Feigl, Herbert. "Mind-Body, *Not* A Pseudoproblem" in *Dimensions of Mind*, ed. S. Hook, pp. 24-36. (First published in 1960) New York: New York University Press, 1964.

Feyerabend, Paul K. "Comment: Mental Events and the Brain", *Journal of Philosophy*, 60 (1963), 295-96.

—. "Materialism and the Mind-Body Problem", *Review of Metaphysics*, 17 (1963), 49-66.

Forrest, Terry. "P-Predicates" in *Epistemology*, ed. A. Stroll, pp. 83-106. New York: Harper and Row, 1967.

Gerard, Ralph W. "What is Memory?" in *Psychobiology*, ed. Scientific American, pp. 126-31, San Francisco: W.H. Freeman, 1967.

Goldberg, Bruce. "The Correspondence Hypothesis", *Philosophical Review*, 67 (1968), 438-54.

Harris, Errol E. "The Neural-Identity Theory and the Person", *International Philosophical Quarterly*, 6 (1966), 515-37.

Hirst, R.J. *The Problems of Perception*. London: Geo. Allen and Unwin, 1959.

Hume, David. *A Treatise of Human Nature*. New York: Dolphin, 1961.

Hungerland, Isabel C. "My Pains and Yours" in *Epistemology*, ed. A. Stroll, pp. 107-128. New York: Harper and Row, 1967.

James, William. *The Principles of Psychology*, 2 Vols. New York: Dover, 1950.

King, Hugh R. "Professor Ryle and *The Concept of Mind*", *Journal of Philosophy*, 48 (1951), 280-96.

Lashley, K.S. *The Neuropsychology of Lashley*, ed. F.A. Beach *et al*. New York: McGraw Hill, 1960.

Leibniz, G. *The Monadology and Other Philosophical Writings* (translated by R. Latta). London: Oxford University Press, 1925.

Malcolm, Norman. "Wittgenstein's *Philosophical Investigations*" in *The Philosophy of Mind*, ed. V.C. Chappell, pp. 74-100. New Jersey: Prentice-Hall, 1962.

—. "Scientific Materialism and the Identity Theory" (Abstract), *Journal of Philosophy*, 60 (1963), 662-63.

Melzack, Ronald. "The Perception of Pain" in *Psychobiology*, ed. Scientific American, pp. 299-307. San Francisco: W.H. Freeman, 1966.

Murdock, G.P. *Our Primitive Contemporaries*. New York: Macmillan, 1934.

Murray, David. "Disembodied Brains", *Proceedings of the Aristotelian Society* (unbound paper) January, 1970, 121-38.

Nagel, Ernest. "The Meaning of Reduction in the Natural Sciences" in *Readings in Philosophy of Science*, ed. P.P. Wiener, pp. 531-49. New York: Scribner, 1953.

Nagel, Ernest and Cohen, Morris. *An Introduction to Logic and Scientific Method*. (First published in 1934) London: Routledge and Kegan Paul, 1964.

Penelhum, Terence. "Personal Identity" in *Encyclopedia of Philosophy*, Vol. 6. ed. Paul Edwards, pp. 95-106. London: Collier-Macmillan, 1967.

Penfield, Wilder. "The Interpretive Cortex", *Science*, 129 (1959), 1719-25.

—. "The Nature of Speech" in *Memory, Learning and Language*, ed. Wm. Feindel, pp. 55-69. Toronto: University of Toronto Press, 1960.

Penfield, Wilder, and Roberts, Lamar. *Speech and Brain-Mechanisms*. Princeton: Princeton University Press, 1959.

Place, U.T. "Materialism as a Scientific Hypothesis", *Philosophical Review*, 69 (1960), 101-04.

—. "Is Consciousness a Brain Process?" in *The Philosophy of Mind*, ed. V.C. Chappell, pp. 101-09. (First published in *British Journal of Psychology*, 1956) New Jersey: Prentice-Hall, 1962.

Popper, Karl R. *The Logic of Scientific Discovery*. New York: Harper and Row, 1965.

Rorty, Richard. "Mind-Body Identity, Privacy, and Categories" in *Philosophy of Mind*, ed. Stuart Hampshire, pp. 30-63. New York: Harper and Row, 1966.

Ryle, Gilbert. *The Concept of Mind*. (First published in 1949) New York: Barnes and Noble, 1962.

—. *Dilemmas*. (First published in 1954) London: Cambridge University Press, 1966.

Shaffer, Jerome. "Could Mental States be Brain Processes?", *Journal of Philosophy*, 58 (1961), 813-22.

—. "Persons and Their Bodies", *Philosophical Review,* 75 (1966), 59-77.

Smart, J.J.C. "Colours", *Philosophy,* 36 (1961), 128-43.

—. "Sensations and Brain Processes" in *The Philosophy of Mind,* ed. V.C. Chappell, pp. 160-72. New Jersey: Prentice-Hall, 1962.

—. *Philosophy and Scientific Realism.* London: Routledge and Kegan Paul, 1963.

—. "Symposium: Materialism", *Journal of Philosophy,* 60 (1963), 651-63.

Smith, Norman Kemp. *New Studies in the Philosophy of Descartes.* London: Macmillan, 1952.

Spinoza, Benedict De. *The Chief Works of Benedict De Spinoza,* 2 Vols., ed. R.H.M. Elwes, New York: Dover, 1955.

Strawson, P.F. *Individuals.* (First published in 1959) London: Methuen, 1965.

Stroll, Avrum. "Statements" in *Epistemology,* ed. A. Stroll, pp. 179-203. New York: Harper and Row, 1967.

Teichmann, Jenny. "The Contingent Identity of Minds and Brains", *Mind,* 76 (1967), 404-15.

Wisdom, John. *Other Minds.* (First published in 1952) Oxford: Blackwell, 1965.

Wittgenstein, Ludwig. *Philosophical Investigations.* New York: Macmillan, 1960.

—. *The Blue and Brown Books.* (Dictated 1934-35) New York: Harper and Row, 1964.

—. *Zettel.* Oxford: Blackwell, 1967.

BIBLIOGRAPHY

Aaron, R.I. "Dispensing with Mind", *Proceedings of the Aristotelian Society,* 52 (1952), 226-42.

Adrian, E.D. "What Happens When We Think" in *The Physical Basis of Mind,* ed. P. Laslett, pp. 5-11, Oxford: Blackwell, 1950.

—. "Consciousness" in *Brain and Conscious Experience,* ed. J.C. Eccles, New York: Springer-Verlag, 1965.

Aldrich, Virgil C. "An Aspect Theory of the Mind", *Philosophy and Phenomenological Research,* 26 (1965-66), 313-26.

Anaxagoras. *The Philosophy of Anaxagoras,* ed. F.M. Cleve. New York: Columbia University Press, 1949.

Anderson, Alan Ross. (ed.), *Minds and Machines.* New Jersey: Prentice-Hall, 1964.

Aristotle. *De Anima—The Works of Aristotle,* ed. W.D. Ross in *Great Books of the Western World,* Vol. 8, ed. R.M. Hutchins, *Encyclopaedia Britannica,* 1952.

Armstrong, D.M. "Is Introspective Knowledge Incorrigible?", *Philosophical Review,* 72 (1963), 417-432.

Ayer, A.J. "The Physical Basis of Mind: A Philosopher's Symposium" in *The Physical Basis of Mind,* ed. P. Laslett, pp. 70-74. Oxford: Blackwell, 1950.

—. *Philosophical Essays.* London: Macmillan, 1954.

—. *Language, Truth and Logic.* (First published in 1936) London: Gollancz, 1958.

—. *The Concept of a Person.* London: Macmillan, 1963.

Baier, Kurt. "Pains", *Australasian Journal of Philosophy,* 40 (1962), 1-23.

—. "Smart on Sensations", *Australasian Journal of Philosophy,* 40 (1962), 57-68.

Barker, S.F. "Book Review of *Concepts, Theories and the Mind-Body Problem,* (eds. Feigl, Scriven and Maxwell. Minneapolis: University of Minnesota Press, 1958)" in *Philosophical Review,* 68 (1959), 391-395.

Barrett, Cyril. (ed.). *Wittgenstein.* Oxford: Blackwell, 1967.

Bedford, Errol. "Emotions" in *The Philosophy of Mind,* ed. V.C. Chappell, pp. 110-126. New Jersey: Prentice-Hall, 1962.

Beloff, John. *The Existence of Mind.* London: MacGibbon and Kee, 1962.

—. "The Identity Hypothesis: A Critique" in *Brain and Mind,* ed. J.R. Smythies, pp. 35-61. London: Routledge and Kegan Paul, 1965.

Black, Max. "The Identity of Indiscernibles", *Mind,* 61 (1952), 153-164.

—. *Philosophy in America.* London: Geo. Allen and Unwin, 1965.

Bok, S.T. "Quantitative Analysis of the Morphological Elements of the Cerebral Cortex" in *Structure and Function of the Cerebral Cortex,* eds. D.B. Tower and J.P. Schade, pp. 7-17. London: Elsevier, 1960.

Boring, Edwin G. *The Physical Dimensions of Consciousness*. London: Century, 1933.

Bosanquet, Bernard. *Essays and Addresses*. London: Swan Sonnenschein, 1889.

Bowman, Andrew. "Knowledge of Other Minds", *Journal of Philosophy*, 50 (1953), 328-32.

Brain, W. Russell. "Some Aspects of the Brain-Mind Relationship" in *Brain and Mind*, ed. J.R. Smythies, pp. 63-80. London: Routledge and Kegan Paul, 1965.

Broad, C.D. *The Mind and Its Place in Nature*. London: Kegan Paul, Trench, Trubner, 1925.

Brodbeck, May. "Discussion: Objectivism and Interaction: A Reaction to Margolis", *Philosophy of Science*, 33 (1966), 287-292.

Burt, Cyril. "Mind and Consciousness" in *The Scientist Speculates*, ed. I.J. Good, pp. 78-79. London: Heinemann, 1962.

Butler, R.J. (ed.). *Analytical Philosophy*. Oxford: Blackwell, 1962.

——. (ed.). *Analytical Philosophy*. Second Series. Oxford: Blackwell, 1965.

Carnap, Rudolf. "Logical Foundations of the Unity of Science" in *Readings in Philosophical Analysis*, eds. H. Feigl and W. Sellars, pp. 408-23. New York: Appleton-Century-Crofts, 1949.

Chappell, V.C. (ed.) *The Philosophy of Mind*. New Jersey: Prentice-Hall, 1962.

Chisholm, Roderick M. "A Book Review of *Mind, Perception and Science* (by W. R. Brain, Oxford: Blackwell, 1951)" in *Journal of Philosophy*, 50 (1953), 503-505.

Chisholm, Roderick M. and Sellars, Wilfrid. "Intentionality and the Mental" in *Minnesota Studies in The Philosophy of Science*, Vol. 2, eds. H. Feigl, M. Scriven, G. Maxwell, pp. 507-33. Minneapolis: University of Minnesota Press, 1958.

Coburn, Robert C. "Shaffer on the Identity of Mental States and Brain Processes", *Journal of Philosophy*, 60 (1963), 89-92.

——. "Pains and Space", *Journal of Philosophy*, 63 (1966), 381-96.

Cohen, Morris R. and Nagel, Ernest. *An Introduction to Logic and Scientific Method*. (First published in 1934) London: Routledge and Kegan Paul, 1964.

Colodny, Robert G. (ed.). *Frontiers of Science and Philosophy*. Pittsburgh: University of Pittsburgh, 1962.

Cornman, James W. "The Identity of Mind and Body", *Journal of Philosophy*, 59 (1962), 486-92.

——. "Mental Terms, Theoretical Terms and Materialism", *Philosophy of Science*, 35 (1968), 45-63.

Culbertson, James T. *The Mind of Robots*. Urbana: University of Illinois Press, 1963.

Delafresnaye, J.F. (ed.) *Brain Mechanisms and Consciousness*. Oxford: Blackwell, 1954.

Dodwell, P.C. "Causes of Behavior and Explanation in Psychology", *Mind*, 69 (1960), 1-13.

Ducasse, C.J. "The Philosophical Importance of 'Psychic Phenomena'", *Journal of Philosophy*, 51 (1954), 810-23.

——. "Minds, Matter and Bodies" in *Brain and Mind*, ed. J.R. Smythies, pp. 81-109. London: Routledge and Kegan Paul, 1965.

Eccles, John C. *The Neurophysiological Basis of Mind*. Oxford: Clarendon, 1953.

Edwards, Paul. (ed.). *The Encyclopedia of Philosophy* (8 Vols.). London: Collier-Macmillan, 1967.

Ellis, Albert. "An Operational Reformulation of Some of the Basic Principles of Psychoanalysis" in *Minnesota Studies in the Philosophy of Science*, Vol. 1. eds. H. Feigl and M. Scriven, pp. 131-154. Minneapolis: University of Minnesota Press, 1956.

Farrell, B.A. "Experience" in *The Philosophy of Mind*, ed. V.C. Chappell, pp. 23-48. (First printed in *Mind*, 1950) New Jersey: Prentice-Hall, 1962.

Feigl, Herbert. "Logical Empiricism" [1] and "Some Remarks on the Meaning of Scientific Explanation" [2] in *Readings in Philosophical Analysis*, eds. H. Feigl and W. Sellars, (1. pp. 3-26; 2. pp. 510-14). New York: Appleton-Century-Crofts, 1949.

—. "The Mind-Body Problem in the Development of Logical Empiricism", *Revue Internationale de Philosophie*, 4 (1950), 64-83.

Feigl, Herbert and Sellars, Wilfrid. (eds.). *Readings in Philosophical Analysis*. New York: Appleton-Century-Crofts, 1949.

Feigl, Herbert and Scriven, Michael. (eds.). *Minnesota Studies in the Philosophy of Science*, Vol. 1. Minneapolis: University of Minnesota Press, 1956.

Feigl, Herbert, Scriven, Michael and Maxwell, Grover. (eds.). *Minnesota Studies in the Philosophy of Science*, Vol. 2. Minneapolis: University of Minnesota Press, 1958.

Feindel, William. "The Brain Considered as a Thinking Machine" in *Memory, Learning and Language*, ed. Wm. Feindel, pp. 11-23. Toronto: University of Toronto Press, 1960.

Fessard, A.E. "Mechanisms of Nervous Integration and Conscious Experience" in *Brain Mechanisms and Consciousness*, ed. J.F. Delafresnaye, pp. 200-36. Oxford: Blackwell, 1954.

Findlay, J.N. "Linguistic Approach to Psychophysics", *Proceedings of the Aristotelian Society*, 50 (1950), 43-64.

Flew, Antony. *Hume's Philosophy of Belief: A Study of His First Inquiry*. London: Routledge and Kegan Paul, 1961.

Fodor, Jerry A. "Explanations in Psychology" in *Philosophy in America*, ed. Max Black, pp. 161-79. London: Geo. Allen and Unwin, 1965.

Fois, Albert. *The Electroencephalogram of the Normal Child* (translated and edited by N.L. Low). Springfield: Thomas, 1961.

Fortuyn, J.D. "The Contributions of Cortex, Telencephalon, Diencephalon and Midbrain to the State of Sleep and Wakefulness" in *Structure and Function of the Cerebral Cortex*, eds. D. B. Tower and J.P. Schade, pp. 88-92. London: Elsevier, 1960.

Frege, Gottlob. "On Sense and Nominatum" in *Reading in Philosophical Analysis*, eds. H. Feigl and W. Sellars, pp. 85-102. New York: Appleton-Century-Crofts, 1949.

Gabor, Dennis. "The Dimensions of Consciousness" in *The Scientist Speculates*, ed. I.J. Good, pp. 66-70. London: Heinemann, 1962.

Geach, Peter. *Mental Acts*. London: Routledge and Kegan Paul, 1957.

Goldberg, Alfred L. "The Death of Dualism", *Harvard Review*, 3 (1965), 24-35.

Good, Irving J. (ed.). *The Scientist Speculates*. London: Heinemann, 1962.

Gunderson, Keith. "The Imitation Game" in *Minds and Machines*, ed. A.R. Anderson, pp. 60-71. New Jersey: Prentice-Hall, 1964.

Hallie, Philip P. "The Privacy of Experience", *Journal of Philosophy,* 58 (1961), 337-46.

Hamlyn, D.W. "Behavior" in *The Philosophy of Mind,* ed. V.C. Chappell, pp. 60-75. (First published in *Philosophy,* 1953) New Jersey: Prentice-Hall, 1962.

Hampshire, Stuart. "The Concept of Mind (Gilbert Ryle)", *Mind,* 59 (1950), 237-55.

—. (ed.). *Philosophy of Mind.* New York: Harper and Row, 1966.

Hanneborg, Knut. *Anthropological Circles.* Copenhagen: Munksgaard, 1962.

Hartnack, Justus. "Remarks on the Concept of Sensation", *Journal of Philosophy,* 56 (1959), 111-17.

Hayek, F.A. *The Sensory Order.* London: Routledge and Kegan Paul, 1952.

Hebb, D.O. "The Problem of Consciousness and Introspection" in *Brain Mechanisms and Consciousness,* ed. J.F. Delafresnaye, pp. 402-21. Oxford: Blackwell, 1954.

Hempel, Carl G. "The Logical Analysis of Psychology" in *Readings in Philosophical Analysis,* eds. H. Feigl and W. Sellars, pp. 373-84. (First published in *Revue de Synthèse,* 1935) New York: Appleton-Century-Crofts, 1949.

Hirst, R.J. "Perception, Science and Common Sense," *Mind,* 60 (1951), 481-505.

—. "Part Three" in *Human Senses and Perception,* ed. G.M. Wyburn, pp. 242-337. Edinburgh: Oliver and Boyd. 1964.

Hobhouse, L.T. *Sociology and Philosophy.* London: G. Bell, 1966.

Hoffer, A. "Modification of Processes of Thought by Chemicals" in *Memory, Learning and Language,* ed. Wm. Feindel, pp. 24-9. Toronto: University of Toronto Press, 1960.

Hofstadter, Albert. "Professor Ryle's Category-Mistake", *Journal of Philosophy,* 48 (1951), 257-70.

Hook, Sidney. (ed.). *Dimensions of Mind.* (First published in 1960) New York: New York University Press, 1964.

Hubel, David H. "The Physiology of the Brain in 1965", *Harvard Review,* 3 (1965), 18-23 .

Ishiguro, Hide. "Imagination" in *British Analytical Philosophy,* eds. B. Willams and A. Montefiore, pp. 153-78. London: Routledge and Kegan Paul, 1966.

Jones, E.E.C. "Precise and Numerical Identity", *Mind,* 17 (1908), 384-393.

Kekes, John. "Physicalism, the Identity Theory and the Doctrine of Emergence", *Philosophy of Science,* 33 (1966), 360-75.

Kneale, Martha. "What is the Mind-Body Problem?", *Proceedings of the Aristotelian Society,* 50 (1950), 105-22.

Kohler, Wolfgang. *The Place of Value in a World of Facts.* New York: Liveright, 1938.

Kubie, L.S. "Psychiatric and Psychoanalytic Considerations of the Problem of Consciousness" in *Brain Mechanisms and Consciousness,* ed. J.F. Delafresnaye, pp. 444-69. Oxford: Blackwell, 1954.

Kuhlenbeck, Hartwig. "The Concept of Consciousness in Neurological Epistemology" in *Brain and Mind,* ed. J.R. Smythies, pp. 137-61. London: Routledge and Kegan Paul, 1965.

Kuiper, John. "Roy Wood Sellars on the Mind-Body Problem", *Philosophy and Phenomenological Research,* 15 (1954), 48-64.

Lachs, John. "Epiphenomenalism and the Notion of Cause", *Journal of Philosophy,* 60 (1963), 141-46.

—. "The Impotent Mind", *Review of Metaphysics,* 17 (1963-64), 187-99.

Laslett, Peter. (ed.) *The Physical Basis of Mind.* Oxford: Blackwell, 1950.

Lean, M.E. "Mr. Gasking on Avowals" in *Analytical Philosophy,* ed. R.J. Butler, Oxford: Blackwell, 1962.

Leibniz, G. *Discourse on Metaphysics,* ed. P.G. Lucas and L. Grint. Manchester: Manchester University Press, 1953.

Lewis, C.I. "Some Logical Considerations Concerning the Mental" in *Readings in Philosophical Analysis,* eds. H. Feigl and W. Sellars, pp. 385-92. (First published in *Journal of Philosophy,* 1944) New York: Appleton-Century-Crofts, 1949.

Lewis, D. "An Argument for The Identity Theory", *Journal of Philosophy,* 63 (1966), 17-25.

Lewis, H.D. "Mind and Body—Some Observations on Mr. Strawson's View", *Proceedings of the Aristotelian Society,* 63 (1962), 1-22.

Libet, B. "Brain Stimulation and the Threshold of Conscious Experience" in *Brain and Conscious Experience,* ed. J.C. Eccles, pp. 165-81. New York: Springer-Verlag, 1965.

Locke, John. *An Essay Concerning Human Understanding.* London: Wm. Baynes (24th edition), 1823.

Lovejoy, Arthur O. *The Revolt Against Dualism.* London: Geo. Allen and Unwin, 1930.

Lucas, J.R. "Minds, Machines and Godel" in *Minds and Machines,* ed. A. R. Anderson, pp. 43-59. New Jersey: Prentice-Hall, 1964.

McConnell, James V. "Memories, Molecules and Minds," *Harvard Review,* 3 (1965), 8-17.

MacIntyre, Alasdair. "The Antecedents of Action" in *British Analytical Philosophy,* eds. B. Williams and A. Montefiore, pp. 205-25. London: Routledge and Kegan Paul, 1966.

MacKay, D.M. "Mentality in Machines", *Aristotelian Society:* Supplementary Volume, 26 (1952), 61-86.

—. "From Mechanism to Mind" in *Brain and Mind,* ed. J.R. Smythies, pp. 163-200. London: Routledge and Kegan Paul, 1965.

Margolis, Joseph. "Objectivism and Interactionism", *Philosophy of Science,* 33 (1966), 118-23.

Mehlberg, Henry. "The Range and Limits of the Scientific Method", *Journal of Philosophy,* 51 (1954), 285-94.

Mill, John Stuart. *A System of Logic,* Vol. 2. London: Longmans, (8th edition), 1872.

Miller, Dickinson S. "Descartes' Myth and Professor Ryle's Fallacy", *Journal of Philosophy,* 48 (1951), 270-280.

Mundle, C.W.K. "Mr. Hirst's Theory of Perception", *Mind,* 61 (1952), 386-90.

—. "Common Sense Versus Mr. Hirst's Theory of Perception", *Proceedings of the Aristotelian Society,* 60 (1959), 61-78.

Nagel, Ernest, *The Structure of Science.* (First published in 1961) London: Routledge and Kegan Paul, 1968.

Negley, Glenn. "Cybernetics and Theories of Mind", *Journal of Philosophy,* 48 (1951), 574-82.

Nochlin, Philip. "Reducibility and Intentional Words", *Journal of Philosophy,* 50 (1953), 625-38.

Olafson, F.A. "A Note on Perceptual Illusion", *Journal of Philosophy,* 50 (1953), 274-8.

O'Shaughnessy, Brian. "The Origin of Pain", *Analysis,* 15 (1955), 121-30. (Reprinted by Blackwell (Oxford), 1967.)

Pap, Arthur. "Semantic Analysis and Psycho-Physical Dualism", *Mind,* 61 (1952), 209-221.

Penfield, Wilder. "The Cerebral Cortex and the Mind of Man" in *The Physical Basis of Mind,* ed. P. Laslett, pp. 56-64. Oxford: Blackwell, 1950.

Pieron, H. *Principles of Experimental Psychology.* London: Kegan Paul, 1929.

Pitcher, George. (ed.). *Wittgenstein: The Philosophical Investigations.* New York: Anchor, 1966.

Pitcher, George and Joske, W.D. "Sensations and Brain Processes: A Reply to Professor Smart", Parts I and II, *Australasian Journal of Philosophy,* 38 (1960), 150-60.

Plato. *The Phaedo of Plato,* ed. B. Jowett. Berkshire: Waltham St. Lawrence, 1930.

Polanyi, Michael. *Personal Knowledge.* London: Routledge and Kegan Paul, 1958.

Popper, Karl. "A Note on the Body-Mind Problem", *Analysis,* 15 (1955), 131-35. (Reprinted by Blackwell (Oxford), 1967.)

Porter, Arthur. "The Mechanical Representation of Processes of Thought" in *Memory, Learning and Language,* ed. Wm. Feindel, pp. 35-54. Toronto: University of Toronto Press, 1960.

Poynter, F.N.L. (ed.). *The Brain and Its Functions: A Symposium.* Oxford: Blackwell, 1958.

Price, H.H. "Seeming", *Aristotelian Society: Supplementary Volume,* 26 (1952), 215-34.

—. "Survival and the Idea of 'Another World' " in *Brain and Mind,* ed. J.R. Smythies, pp. 1-24. London: Routledge and Kegan Paul, 1965.

Putnam, Hilary. "A Review of Patterns of Discovery, by N.R. Hanson" (Cambridge, 1958), *Science,* 129 (1959), 1666-67.

—. "Dreaming and 'Depth Grammar' " in *Analytical Philosophy,* ed. R.J. Butler, pp. 210-35. Oxford: Blackwell, 1962.

—. "Brains and Behavior" in *Analytical Philosophy, Second Series,* ed. R.J. Butler, pp. 1-19. Oxford: Blackwell, 1965.

Quinton, A.M. "Seeming", *Aristotelian Society: Supplementary Volume,* 26 (1952), 235-52.

—. "Mind and Matter" in *Brain and Mind,* ed. J.R. Smythies, pp. 201-39. London: Routledge and Kegan Paul, 1965.

Raab, Francis V. "Of Minds and Molecules", *Philosophy of Science,* 32 (1965), 57-72.

Rhine, J.B. "The Science of Nonphysical Nature", *Journal of Philosophy,* 51 (1954), 801-810.

Roelofs, Howard D. "A Case for Dualism and Interaction", *Philosophy and Phenomenological Research,* 15 (1955), 451-76.

Russel, Bertrand. *The Principles of Mathematics,* Vol. 1. London: Cambridge University Press, 1903.

—. *The Philosophy of Leibniz.* London: Geo. Allen and Unwin (2nd edition), 1937.

—. "What is Mind?", *Journal of Philosophy*, 55 (1958), 5-12.

Russell, Bertrand and Whitehead, A.N. *Principia Mathematica*, Vol. 1. London: Cambridge University Press (2nd edition), 1925.

Samuel, P.C. "The Physical Basis of Mind: A Philosopher's Symposium" in *The Physical Basis of Mind*, ed. P. Laslett, pp. 65-69. Oxford: Blackwell, 1950.

Savery, Barnett. "Identity and Difference", *Philosophical Review*, 51 (1942), 205-212.

Schilpp, Paul. (ed.). *The Philosophy of C.D. Broad*. New York: Tudor, 1959.

Schlick, Moritz. "On the Relation Between Psychological and Physical Concepts" in *Readings in Philosophical Analysis*, eds. H. Feigl and W. Sellars, pp. 393-407. (First published in *Revue de Synthèse*, 1935) New York: Appleton-Century-Crofts, 1949.

Schmitt, Francis O. "Molecules and Memory", *Harvard Review*, 3 (1965), 1-17.

Schrodinger, Erwin. *Mind and Matter*. London: Cambridge University Press, 1958.

Scriven, Michael. "A Study of Radical Behaviorism" in *Minnesota Studies in the Philosophy of Science*, Vol. 1, eds. H. Feigl and M. Scriven, pp. 89-130. Minneapolis: University of Minnesota Press, 1956.

—. "The Mechanical Concept of Mind" in *Minds and Machines*, ed. A. R. Anderson, pp. 31-42. New Jersey: Prentice-Hall, 1964.

—. *Primary Philosophy*. New York: McGraw Hill, 1966.

Scriven, Michael and Feigl, Herbert. (eds.). *Minnesota Studies in the Philosophy of Science*, Vol. 1. Minneapolis: University of Minnesota Press, 1956.

Scriven, Michael, Feigl, Herbert and Maxwell, G. (eds.). *Minnesota Studies in the Philosophy of Science*, Vol. 2. Minneapolis: University of Minnesota Press, 1958.

Sellars, Wilfrid. "Realism and the New Way of Words" in *Readings in Philosophical Analysis*, eds. H. Feigl and W. Sellars, pp. 424-56. New York: Appleton-Century-Crofts, 1949.

—. "Empiricism and the Philosophy of Mind" in *Minnesota Studies in the Philosophy of Science*, Vol. 1, eds. H. Feigl and M. Scriven, pp. 253-329. Minneapolis: University of Minnesota Press, 1956.

—. "Philosophy and the Scientific Image of Man" in *Frontiers of Science and Philosophy*, ed. R.G. Colodny, pp. 35-78. Pittsburgh: University of Pittsburgh, 1962.

Sellars, Wilfrid and Feigl, Herbert. (eds.). *Readings in Philosophical Analysis*. New York: Appleton-Century-Crofts, 1949.

Sellars, Wilfrid and Chisholm R. "Intentionality and the Mental" in *Minnesota Studies in the Philosophy of Science*, Vol. 2, eds. H. Feigl, M. Scriven and G. Maxwell, pp. 507-533. Minneapolis: University of Minnesota Press, 1958.

Shaffer, Jerome. "Mental Events and the Brain", *Journal of Philosophy*, 60 (1963), 160-66.

Sherrington, Charles. *Man on His Nature*. London: Cambridge University Press, 1940.

Sinnott, Edmund W. *Cell and Psyche*. (First published University of North Carolina Press, 1950) New York: Harper, 1961.

Smart, J.J.C. "Ryle on Mechanism and Psychology", *Philosophical Quarterly*, 9 (1959), 349-55.

—. "Sensations and Brain Processes", *Philosophical Review*, 68 (1959), 141-56. (Reprinted in Chappell, 1962 with some variations.)

—. "Sensations and Brain Processes: A Rejoinder to Dr. Pitcher and Mr. Joske", *Australasian Journal of Philosophy*, 38 (1960), 252-54.

—. "Brain Processes and Incorrigibility", *Australasian Journal of Philosophy*, 40 (1962), 68-70.

—. "Professor Ziff on Robots" in *Minds and Machines*, ed. A.R. Anderson, pp. 104-5. (First published in *Analysis*, 1959) New Jersey: Prentice-Hall, 1964.

Smart, Ninian. "Robots Incorporated" in *Minds and Machines*, ed. A.R. Anderson, pp. 106-8. (First published in *Analysis*, 1959) New Jersey: Prentice-Hall, 1964.

Smythies, J.R. "The Representative Theory of Perception" in *Brain and Mind*, ed. J.R. Smythies, pp. 241-64. London: Routledge and Kegan Paul, 1965.

Stace, W.T. "The Refutation of Realism" in *Readings in Philosophical Analysis*, eds. H. Feigl and W. Sellars, pp. 364-72. (First published in *Mind*, 1934) New York: Appleton-Century-Crofts, 1949.

Stevenson, J.T. "Sensations and Brain Processes: A Reply to J.J.C. Smart", *Philosophical Review*, 69 (1960), 505-10.

Strawson, P.F. "Persons" in *Minnesota Studies in the Philosophy of Science*, Vol. 2, eds. H. Feigl, M. Scriven and G. Maxwell, pp. 330-53. Minneapolis: University of Minnesota Press, 1958.

Stroll, Avrum. (ed.). *Epistemology: New Essays in the Theory of Knowledge*. New York: Harper and Row, 1967.

Strong, C.A. *Why the Mind Has a Body*. London: Macmillan, 1903.

Teuber, H.L. "Convergences, Divergences, Lacunae" in *Brain and Conscious Experience*, ed. J.C. Eccles, pp. 575-83. New York: Springer-Verlag, 1965.

Thomson, J.F. "Reducibility", *Aristotelian Society: Supplementary Volume*, 26 (1952), 87-104.

Tucker, John. "A Review of Hirst's *The Problems of Perception*" (London: Geo. Allen and Unwin, 1959), *Mind*, 69 (1960), 596-71.

Turing, A.M. "Computing Machinery and Intelligence" in *Minds and Machines*, ed. A.R. Anderson, pp. 4-30. New Jersey: Prentice-Hall, 1964.

Van Peursen, C.A. *Body, Soul, Spirit*. London: Oxford University Press, 1966.

Vesey, G.N.A. "The Location of Bodily Sensations", *Mind*, 70 (1961), 25-35.

—. *The Embodied Mind*. London, Geo. Allen and Unwin, 1965.

Warnock, G.J. "Reducibility", *Aristotelian Society: Supplementary Volume*, 26 (1952), 105-20.

Weil, Andrew T. "Introduction — The New Psychology", *Harvard Review*, 3 (1965), i-viii.

Weiss, E.H. "A Review of *Mechanical Man* by Dean E. Wooldridge" (New York: McGraw, 1968), *Review of Metaphysics*, 21 (1968), 758.

Weissman, Asriel. "The Meaning of Identity", *Philosophy and Phenomenological Research*, 16 (1955-56), 461-75.

Weitz, Morris. "Professor Ryle's 'Logical Behaviorism'", *Journal of Philosophy*, 48 (1951), 297-301.

Whitehead, Alfred North. *Process and Reality*. London: Cambridge University Press, 1929.

Whiteley, C.H. "A Note on the Concept of Mind", *Analysis,* 16 (1956), 68-70. (Reprinted by Blackwell (Oxford), 1967.)

Wiener, Philip P. (ed.). *Readings in Philosophy of Science.* New York: Scribner, 1953.

Wiggins, D. "Identity Statements" in *Analytical Philosophy,* Second Series, ed. R.J. Butler, pp. 40-71. Oxford: Blackwell, 1965.

Wigner, Eugene P. "Remarks on the Mind-Body Question" in *The Scientist Speculates,* ed. I.J. Good, pp. 284-302. London: Heinemann, 1962.

Wilkie, J.S. *The Science of Mind and Brain.* London: Hutchinson's University Library, 1953.

Williams, B.A.O. "Mr. Strawson on Individuals", *Philosophy,* 36 (1961), 309-32.

Williams, B.A.O. and Montefiore, Alan. (eds.). *British Analytical Philosophy.* London. Routledge and Kegan Paul, 1966.

Williams, Denis. "Old and New Concepts of the Basis of Consciousness" in *The Brain and its Functions: A Symposium,* ed. F.N.L. Poynter, pp. 73-82. Oxford: Blackwell, 1958.

Wisdom, John. *Problems of Mind and Matter.* (First published in 1934) London: Cambridge University Press, 1963.

—. *Philosophy and Psychoanalysis.* (First published in 1953) Oxford: Blackwell, 1964.

—. *Paradox and Discovery.* Oxford: Blackwell, 1965.

Wisdom, J.O. "Some Main Mind-Body Problems", *Proceedings of the Aristotelian Society,* 60 (1960), 187-210.

Wittgenstein, Ludwig. *Tractatus Logico-Philosophicus.* London: Kegan Paul, Trench, Trubner, 1922.

Wolman, Benjamin B. "Principles of Monistic Transitionism" in *Scientific Psychology,* ed. B.B.Wolman, pp. 563-85. London: Basic Books, 1965.

Wooldridge, Dean E. *Mechanical Man.* New York: McGraw-Hill, 1968.

Wright, J.N. "Mind and the Concept of Mind", *Aristotelian Society: Supplementary Volume,* 33 (1959), 1-22.

Yolton, John W. "The Dualism of Mind", *Journal of Philosophy,* 51 (1954), 173-80.

—. "Book Review — *Perception and the Physical World,* D.M. Armstrong, (New York: Humanities Press, 1961)" *Journal of Philosophy,* 60 (1963), 384-88.

Young, J.Z.. "Summary Statement Neuroanatomy Symposium" in *Structure and Function of the Cerebral Cortex,* eds. D.B. Tower and J.P. Schade, pp. 165-70. London: Elsevier, 1960.

Ziff, Paul. "About Behaviorism" in *The Philosophy of Mind,* ed. V.C. Chappell, pp. 147-150. New Jersey: Prentice-Hall, 1962.

—. "The Feelings of Robots" in *Minds and Machines,* ed. A.R. Anderson, pp. 98-103. New Jersey: Prentice-Hall, 1964.

INDEX